Irena Zupanič Pajnič

Genetic Identification of Second World War Victim's Skeletal Remains

I0431965

Irena Zupanič Pajnič

Genetic Identification of Second World War Victim's Skeletal Remains

LAP LAMBERT Academic Publishing

Impressum / Imprint

Bibliografische Information der Deutschen Nationalbibliothek: Die Deutsche Nationalbibliothek verzeichnet diese Publikation in der Deutschen Nationalbibliografie; detaillierte bibliografische Daten sind im Internet über http://dnb.d-nb.de abrufbar.

Alle in diesem Buch genannten Marken und Produktnamen unterliegen warenzeichen-, marken- oder patentrechtlichem Schutz bzw. sind Warenzeichen oder eingetragene Warenzeichen der jeweiligen Inhaber. Die Wiedergabe von Marken, Produktnamen, Gebrauchsnamen, Handelsnamen, Warenbezeichnungen u.s.w. in diesem Werk berechtigt auch ohne besondere Kennzeichnung nicht zu der Annahme, dass solche Namen im Sinne der Warenzeichen- und Markenschutzgesetzgebung als frei zu betrachten wären und daher von jedermann benutzt werden dürften.

Bibliographic information published by the Deutsche Nationalbibliothek: The Deutsche Nationalbibliothek lists this publication in the Deutsche Nationalbibliografie; detailed bibliographic data are available in the Internet at http://dnb.d-nb.de.

Any brand names and product names mentioned in this book are subject to trademark, brand or patent protection and are trademarks or registered trademarks of their respective holders. The use of brand names, product names, common names, trade names, product descriptions etc. even without a particular marking in this works is in no way to be construed to mean that such names may be regarded as unrestricted in respect of trademark and brand protection legislation and could thus be used by anyone.

Coverbild / Cover image: www.ingimage.com

Verlag / Publisher:
LAP LAMBERT Academic Publishing
ist ein Imprint der / is a trademark of
AV Akademikerverlag GmbH & Co. KG
Heinrich-Böcking-Str. 6-8, 66121 Saarbrücken, Deutschland / Germany
Email: info@lap-publishing.com

Herstellung: siehe letzte Seite /
Printed at: see last page
ISBN: 978-3-659-45306-9

Copyright © 2013 AV Akademikerverlag GmbH & Co. KG
Alle Rechte vorbehalten. / All rights reserved. Saarbrücken 2013

TABLE OF CONTENTS

INTRODUCTION.. 3

NATURE OF ANCIENT DNA.. 11

GENETIC MARKERS USED FOR HUMAN IDENTIFICATION......... 12

SKELETONISED HUMAN REMAINS.. 15
TEETH.. 15
BONES... 18

MEASURES FOR PREVENTING DNA CONTAMINATION.............. 22

EXCAVATION AND STORAGE OF SKELETAL REMAINS..............
27

ELIMINATION DATABASE... 29

FAMILY REFERENCES – RELATIVES OF THE VICTIMS.............. 29

EXTRACTION OF GENOMIC DNA... 31

DNA QUANTIFICATION... 35

DNA TYPING.. 37
AUTOSOMAL STR DNA TYPING.. 38
Y CHROMOSOMAL STR DNA TYPING.. 43
MITOCHONDRIAL DNA TYPING.. 46

STATISTICAL ANALYSES OF FAMILIAR RELATIONSHIPS.......... 47

EXAMPLES OF IDENTIFICATION OF THE SECOND WORLD WAR MASS GRAVE VICTIMS………………………………………………….. 49

EXAMPLE OF GENETIC ANALYSES OF HUMAN SKELETAL REMAINS FROM ARCHAEOLOGICAL SITE……………………….. 53

CONCLUSION………………………………………………………… 57

ACKNOWLEDGEMENTS……………………………………………. 58

LITERATURE………………………………………………………….. 59

INTRODUCTION

In cases where unidentified skeletonised human remains are found and identification cannot be performed using classical forensic medicine methods, bones or teeth can be used for molecular genetic identification. The condition of the skeletal remains analysed for forensic identification studies is often not ideal for DNA recovery. In old bones and teeth, small amounts of endogenous DNA, the presence of polymerase chain reaction (PCR) inhibitors, the degradation of the DNA and the exceptional risk of contamination limit the success of DNA typing (Alaeddini et al. 2010; Alaeddini et al. 2012; Lee et al. 2010a). DNA typing using bone and tooth samples has been successful in forensic identification analysis and anthropological studies (Anderung et al. 2008). It has also become a valuable tool for identifying victims in mass graves and individual graves from the Second World War. Due to the rather long time span since the Second World War, it is difficult to find living relatives to identify the victims in mass graves. If there are no close relatives, more distant relatives can also be very helpful because a combination of genetic markers may provide satisfactory probabilities of identity.

Mitochondrial DNA (mtDNA) testing is regularly employed in forensic identification of aged skeletal remains (Anslinger et al. 2001; Stone et al. 2001; Palo et al. 2007), but mtDNA typing alone is often insufficient for identification, and the analysis of nuclear short tandem repeat (STR) loci is required (Biesecker et al. 2005). Actually, nuclear DNA is the preferred genome of amplification for forensic purposes, as it is individually specific and provides bi-parental kinship information (Lee et al. 2010b). Recently, some researchers (among them is also our group) report the successful typing of nuclear STRs from ancient material (Irwin et al. 2007a; Irwin et al. 2007b; Zupanič Pajnič et al. 2010; Lee et al.

2010c; Vanek et al. 2009; Bogdanowicz et al. 2009). We managed to obtain nuclear DNA for successful STR typing from skeletal remains excavated from the Auersperg chapel archaeological site that were over 300 years old (Zupanič Pajnič 2013a), and we successfully identified victims of massacres that took place during and after the Second World War in Slovenia (Zupanič Pajnič 2008; Zupanič Pajnič et al. 2010). In identifying Second World War victims, we obtained mtDNA haplotypes from the HVI/HVII regions, Y-STR haplotypes and autosomal STR profiles from the bones, which made it possible to compare both close and distant relatives in both the maternal and paternal lines.

The laboratory of Molecular Genetics at the Institute of Forensic Medicine, Faculty of Medicine, University of Ljubljana was established in 1996 and we started with molecular genetic analyses of ancient DNA in 2005. The laboratory is equipped with modern molecular genetic analyses machines (Biorobot EZ1-Qiagen, 7500 Real Time PCR System with HID Real-Time PCR Analysis Software v. 1.1 (Applied Biosystems), ABI PRISM^{TM} 3130 Genetic Analyser with Data Collection v. 3.0 Software and GeneMapper ID v. 3.2 Software (Applied Biosystems)), safety hoods and organization of the rooms in accordance with preventing contamination when typing old skeletal remains using forensic human identification methods. We have a special room for cleaning, cutting and grinding of bones and teeth.

In Figure 1 there are the photos of the Laboratory of Molecular Genetics at the Institute of Forensic Medicine, Faculty of Medicine, University of Ljubljana.

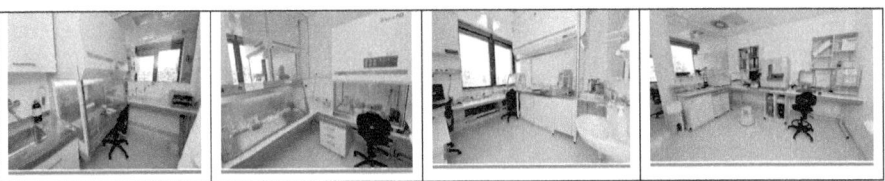

Figure 1: Photos of the Laboratory of Molecular Genetics at the Institute of Forensic Medicine, Faculty of Medicine, University of Ljubljana (photo: Katarina Podgoršek).

Molecular genetic identification of skeletal remains will be presented on an example of victims of Second World War killings in Slovenia. The Commission on Concealed Mass Graves assigned by the Government of the Republic of Slovenia has recently registered more than 600 hidden mass graves from this period (Ferenc 2008). In Figure 2 there are maps of Slovenia where Second World War mass graves are marked. On the left map, all registered Second World War mass graves are marked, and on the right map the mass graves are marked according to the nationality of the victims.

Figure 2: Maps of Slovenia where Second World War mass graves are marked. On the left map, all registered Second World War mass graves are marked, and on the right map the mass graves are marked according to the nationality of the victims (Ferenc 2008).

For some Second World War mass graves in Slovenia, excavation was performed, and in Figure 3 some photos from mass graves are shown.

Figure 3: Photos from Slovenian mass graves (photo: Pavel Jamnik, Jože Dežman, Mehmedalija Alić, Jože Jagrič, Andrej Mihevc).

There is no exact data on the number of Yugoslav communist armed forces victims in the graves, but according to historian Jože Dežman the number of missing persons could be as high as 100,000. For most mass graves, there are no documents (the list of victims with their names) to base victim identification on. However, some of the mass graves represent a rare exception; the largest one among them is the Konfin I mass grave.

At the Laboratory of Molecular Genetics at the Institute of Forensic Medicine, we genetically analysed and positively identified the victims of mass graves where we could access lists of victims, based on which we were able to collect comparative samples of buccal swabs from living relatives. Skeletal remains were analysed for the Konfin I mass grave, located in a karst cave, where 88 victims were killed, the karst cave Konfin II mass grave with 62 excavated skeletons and three mass graves found in the Storžič forest, Bodovlje gorge (Bodoveljska grapa) and Mozelj where massacre victims were excavated from the soil. The skeletal remains were excavated under the direction of local archaeologists and anthropologists. We obtained informed consent from the living relatives for molecular genetic typing of bones and teeth. We identified 32 victims among the skeletal remains from the Konfin I mass grave (Zupanič Pajnič et al. 2010), three victims among the skeletal remains from the mass grave at Mount Storžič (Zupanič Pajnič 2008) and seven victims from the Bodovlje gorge (Bodoveljska grapa) mass grave (Zupanič Pajnič 2007). We matched all victims to living relatives with a posterior probability higher than 99.9% as recommended by Biesecker et al. (2005), Brenner (2003) and Prinz et al. (2007).

Let's look at the biggest mass grave from which we managed to identify victims. For the Konfin I mass grave, a list of the victims (88 Slovenian men taken from the Central Prison on the night of 24 June 1945 and brought to the execution site at Konfin I Cave) can be made based on archives (i.e. the prisoners' logbook of the Yugoslav secret police (OZNA) Central Prison and the registry of detainees with a list of wounded and patients). Among the victims were 40 wounded men and patients that had been transferred from the general hospital in Ljubljana to the OZNA Central Prison 14 days prior to execution and 48 men selected from among the prisoners. These men were not tried in a court and had not been convicted of any crime (Jamnik 2008). Their bodies were thrown into a 45-m-deep karst cave, and the entrance was dynamited. The bodies were not covered with earth that would have kept the skeletons in their original position. Water runoff ran unhindered into the cave, and its 20 m2 bottom was completely covered with a 2 m thick layer of mixed bones filled with mud.

In Figure 4 there are the photos of the karst cave Konfin I from outside (left) and from inside (right).

Figure 4: Photos of the karst cave Konfin I from outside (left) (photo: Mitja Ferenc) and from inside (right) (photo: Pavel Jamnik).

The skeletal remains were excavated under the leadership of a local archaeologist and anthropologist.

In Figure 5 there are photos of excavation of the skeletal remains from the Konfin I mass grave.

Figure 5: The photos of excavation of the skeletal remains from the Konfin I mass grave (photo: Pavel Jamnik - left and Žandi Dežman - right).

An excavation according to **anatomic position** was not possible, and the bones were sorted anatomically by the anthropologist. Through the anthropological study, the number, gender and age of the victims were determined. The anthropologist found that all of the victims were male and that the number of the victims is between 85 and 89.

In Figure 6 there are photos of the Konfin I mass grave bottom, which was completely covered with a 2 m thick layer of mixed bones filled with mud (left), and excavated bones sorted anatomically by the anthropologist (right).

Figure 6: Photos of the Konfin I mass grave bottom, which was completely covered with a 2 m thick layer of mixed bones filled with mud (left) (photo: Stanko Gruden), and excavated bones sorted anatomically by the anthropologist (right) (photo: Pavel Jamnik).

Under a decree by the government of the Republic of Slovenia, the Commission on Concealed Mass Graves in Slovenia entrusted identification of the victims in the Konfin I mass grave to our institute. Thus, after excavation and anthropological study, the bones were sent to us to perform molecular genetic identification. DNA typing involved all excavated right femurs (67 complete and 17 proximal fragments). We collected buccal swabs from family references (sisters, brothers, daughters, sons, wives, cousins and nephews) that were close or distant relatives of 44 Konfin I massacre victims. The bone samples for DNA analysis were collected, labelled and photo documented.

In Figure 7 there are photos of a complete femur (middle), fragments of femurs (left) and sampled bone fragments for genetic analyses (right).

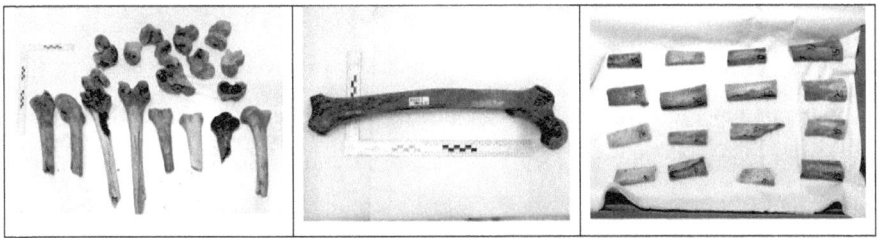

Figure 7: Photos of a complete femur (middle), fragments of femurs (left) and sampled bone fragment for genetic analyses (right) (photo: Rudi Bevc).

NATURE OF ANCIENT DNA

In a living cell, enzyme processes ensure that the DNA stays intact (Lindahl 1993). The balance in the cell gets destroyed after the death of the organism; DNA becomes exposed to degradation by lysosome nucleases and degradation by bacteria and fungi that feed on macromolecules (Eglinton and Logan 1991).

Upon DNA binding to the mineral base of bones and teeth, the enzyme and microbial degradation slows down. Consequently, DNA can be preserved even in exceptionally old skeletal samples but is frequently degraded into fragments of which the length is between 100 and 500 bp (Hofreiter et al. 2001; Pääbo 1989). Preservation of DNA is reduced with age; however, that is not always the case, as the environments surrounding the skeletal samples have the biggest effect on their preservation (Pääbo 1989; Höss et al. 1996; Poinar et al. 1996). The most significant environmental factors are temperature, humidity, pH, chemical characteristics of the soil and the presence of microorganisms. Fast drying out of human remains provides better preservation of DNA; another important factor for its preservation is minimal exposure to UV radiation. DNA can be well preserved in soil rich in salt, with neutral or slightly alkaline pH, low content of humid acids and low humidity (Lindahl 1993; Hagelberg et al. 1991; Tuross 1994). The key factor for DNA preservation is definitely the ambient temperature in which the skeletal remains were located since the time of the organism's death until their exhumation or until molecular genetic testing. The most thoroughly preserved DNA can be found in samples located in caves or in permanently frozen ground - permafrost, as average annual low temperature provides the best possible preservation (Smith et al. 2001; Smith et al. 2003). Another aspect that affects the quality and quantity of DNA in skeletal remains is the storage method used after their exhumation (Burger et al. 1999; Pruvost et al. 2007). Effectiveness of genetic typing is much higher with freshly exhumed skeletons rather than with skeletons that have been kept at room temperature for several years (for example: museum collections), particularly because of higher ambient temperature and washing of the skeletal remains before storing them, which lowers their pH and the concentration of the salt content in the bones. Freezing the skeletal remains is preferred in order to ensure the best preservation of the DNA (Malmstrom 2007). Long bones and teeth are the most appropriate samples for molecular genetic testing, as the DNA in them can stay well preserved for a long time (Miloš et al. 2007; Misner et al. 2009; Edson et al. 2004).

GENETIC MARKERS USED FOR HUMAN IDENTIFICATION

Due to highly degraded DNA acquired from skeletal remains, molecular genetic analyses were based on very short polymorphic fragments of nuclear and mitochondrial DNA (mtDNA). In nuclear DNA, polymerase chain reaction (PCR)-based microsatellites or short tandem repeat (STR) loci are analysed. STRs are typically in the non-coding intron regions and do not code for proteins. The length polymorphisms in STRs are due to the different numbers of copies of the repeat sequence (normally of length 2-6 base pairs) that can occur in a population and the number of copies varies between different individuals. When typing multiple autosomal STRs, unique DNA profiles are obtained and discrimination between unrelated and even closely related individuals is possible. STRs with higher powers of discrimination, tetranucleotide repeat sequences and lengths up to 400 base pairs are chosen for human identification in forensics genetics. Together with autosomal STRs, short fragments of amelogenin gene on X and Y chromosomes are amplified and gender is determined (Zupanič Pajnič 2011b). Apart from autosomal STRs and STRs linked to Y chromosome, we also apply a process of human identification analyses of sequence polymorphisms of the mtDNA control region, which is of central importance for forensic identity testing. Mitochondrial DNA is found in the cell in many copies, which provides extended preservation in the old skeletal remains as opposed to nuclear DNA (Pääbo 1989; Noonan et al. 2006; Valdiosera et al. 2006; Allaeddini et al. 2010). In metabolically active cells, 1000 to 10,000 molecules of mtDNA can be found. Mitochondrial DNA is less likely to become degraded than nuclear DNA, as its circular conformation and mitochondrial membrane keep it from being affected by exonuclease enzyme degradation (Hopwood et al. 1996). Human mtDNA is inherited from the mother and is transferred to the offspring regardless of their sex, which enables us to trace the maternal line, as opposed to the Y chromosome, which is inherited from the father and lets us trace the paternal line. Due to the absence of recombination and inheritance from the mother and the father, individuals with the same nucleotide sequences of mtDNA have a common female ancestor (Tully et al. 2001) and individuals with the same Y chromosome haplotypes have a common male ancestor. This is the base for the identification of skeletal remains through analyses of mtDNA and Y chromosome. The level of identification that we reach using mtDNA and Y chromosome analyses is much lower than that which we achieve through analysis of the microsatellites of autosomal nuclear DNA, due to the absence of recombination. By typing mtDNA and Y chromosomes, we can only identify maternal or paternal line, while the microsatellites of autosomal nuclear DNA allow actual individualisation of the person. That is why in identifying the Second World War victims, we investigate not only polymorphisms of mtDNA, which are the most appropriate to analyse in compromised old bone and tooth DNA samples, but also polymorphisms of nuclear DNA (autosomes and Y chromosome), if possible.

In the process of DNA typing, we encounter not only highly degraded DNA but also very small amounts of endogenous DNA, which are difficult to differentiate from the far more common modern (exogenous) DNA. Unfortunately, contamination is a serious problem in investigations of DNA obtained from old skeletal remains (Hofreiter et al. 2001; Hand et al. 1994; Kolmann and Tuross 2000; Wandeler et al. 2003; Handt et al. 1991). Contamination of the endogenous DNA of bones and teeth with modern DNA can occur during exhumation, improper storage of the skeletal remains and anthropological and molecular genetics investigations (Brown and Brown 1992). Therefore, it is necessary to consider the recommendations to prevent contamination (Pääbo 1990) and the criteria that confirm the authenticity of acquired genetic profiles (Pääbo 1989). It is very important to test the extraction and amplification negative control in parallel with the samples of bones and teeth, along with the staff that participated in the exhumation, storage and anthropological and genetic investigations; doing so allows us to trace contamination in the event of its occurrence. At least two samples must be typed for each skeleton, and it is necessary to obtain identical genetic profiles from both of them (Pääbo et al. 2004).

SKELETONISED HUMAN REMAINS

TEETH

In comparison to bones, DNA in teeth is far more protected. The hard dental tissues that surround the pulp cavity represent a physical protection of the dental pulp (Pfeiffer et al. 1999). Hard dental tissues are preserved even in numerous extreme ambient environments, as they do not undergo immediate post-burial degradation or water dissolution and can withstand temperatures up to 1100°C (Sweet and Hildebrand 1998). Alvarez Garcia et al. (1996) studied the influence of the environment on the degradation of DNA in the teeth and concluded that the fastest degradation of DNA occurs in water, slightly less in soil and least in air. DNA is exceptionally stable in old teeth as it is, similarly to the old bones, bounded to hydroxyapatite (Ohira et al. 1999).

The tooth pulp represents the richest source of DNA in the teeth. It is well protected in the pulp cavity as long as the tooth is firmly anchored in the alveolar bone. If the tooth is stored in a dry environment, dehydration and mummification of the dental pulp occurs, which prevents decay and necrotic processes and allows good preservation of DNA. If the suspension apparatus of the tooth loosens and the microenvironment becomes humid, pulp quickly rots. So 3 weeks after the burial of a tooth in the ground, Schwartz et al. (1991) were no longer able to obtain the nuclear profile due to the destruction of dental pulp; Pfeiffer et al. (1999), however, found a 90% reduction in the quantity of nuclear DNA after six weeks in the ground. In cases where the post-mortem interval is long and the probability of degradation of dental pulp is high, mtDNA typing can be done, of which dentin is the main source - the remains of odontoblast extensions in the dentinal canals (Pfeiffer et al. 1998) - as well as cementites in cement (Smith et al. 1993). Maximum DNA is obtained from the tooth if we grind the whole tooth. That way we are able to capture DNA located in the hard dental tissues (Sweet and Hildebrand 1998). The amount of DNA depends on the size of the dental pulp and type of teeth; the molars are the richest source of DNA. For identifying victims of mass graves, Boles et al. (1995) recommended the use of teeth, as they are easier to transport than the bones, leading to high-quality genomic DNA. Teeth suitable for DNA isolation appear in the following order: endodontically untreated molar, premolar, canine and incisor and endodontically treated molar, premolar, canine and incisor (Budowle et al. 2000).

Mörnstad et al. (1999) isolated DNA from dental pulp and dentin and tried to identify the impact of chronological age on the amount of mtDNA in the dentin in two separate procedures. He noticed a decrease in the amount of mtDNA with age, which is in accordance with degenerative age-related changes in cellular extensions of odontoblasts and the closures of dentinal canals with crystals of calcium phosphate. However, after degradation of the extensions of odontoblasts, a sufficient amount of mtDNA remains among the dentin crystals of calcium phosphate for successful molecular genetic identification. The amount of DNA in endodontically treated – a filled tooth is 100 times lower than in a vital tooth but high enough to successfully amplify in the polymerase chain reaction - PCR. The presence of DNA in the endodontically treated tooth indicates the presence of DNA in the hard tissues (dentin cement) of the tooth, as the pulp was removed. The presence of tooth decay and filling materials do not affect DNA extraction and amplification, so the filling materials do not have to be removed before grinding of the tooth (Sweet and Hildebrand 1998). In the event of explosions or air accidents, it is often necessary to carry out molecular genetic identification on fragmented teeth.

Gaytmenn and Sweet (2003) did research on which part of the tooth contains the highest amount of DNA. They divided the tooth into four sections: the crown body and the crown tip as well as the apical one third and cervical two thirds of the root. They concluded that the highest concentration of DNA is in the cervical two-thirds of the root (dental pulp, dentin and cementum), followed by the cervical part of the crown (dental pulp, dentin and enamel) and the apical third of the root (dental pulp, dentin and cementum). The smallest amount of DNA can be found in the upper section of the crown (dentin, enamel and perhaps diverticula of the dental pulp). The sheer amount of DNA among individual and between groups of teeth varies strongly; quality and quantity of isolated DNA also depend on the tooth pathology, previous dental procedures, elapsed time since the extraction of the tooth until the isolation of DNA and the donor's age (Schwartz et al. 1991). Tsuchimochi et al. (2002) successfully isolated genomic DNA from teeth (dental pulp) that were exposed to a temperature of 300°C for two minutes using the Chelex method, which provided very good results in the identification of charred-burned corpses as well. When the tooth was exposed to a temperature of 500°C or higher, carbonation of the dental pulp occurred, and it lead to the destruction of DNA. Sweet and Sweet (1995) acquired 1.35 µg of DNA out of an impacted wisdom tooth that was found in a charred corpse. Sweet and Hildebrand (1998) extracted from 0.5 up to 97.5 µg of DNA (an average of 30 µg) out of freshly extracted and undamaged molars that they ground finely; on the other hand, Schwartz et al. (1991) extracted from 15 up to 20 µg of DNA. Sweet et al. (1999) obtained a microsatellite nuclear profile of the tooth of the corpse that was exhumed 3.5 years after burial, and Alvarez Garcia et al. (1996) managed to do the same with teeth that were up to 50 years old. Boles et al. (1995) obtained 10-50 µg of DNA from each tooth of several corpses exhumed 10 years after their burial, and successfully typed mtDNA. Baker et al. (2001) obtained mtDNA from teeth that were 3000 years old. We managed to obtain up to 10.7 ng of DNA/g of tooth powder and get complete nuclear profiles of the teeth excavated in Auersperg's chapel at the Ljubljana market that were over 300 years old (Zupanič Pajnič et al. 2012b; Zupanič Pajnič 2013a).

BONES

In the bone tissue, DNA is located in the osteocytes, which are coated with mineralised bone mass. There are between 20,000 and 26,000 osteocytes in 1 mm³ of bone (Hochmeister et al. 1991) Bone tissue is a specialized form of connective tissue composed of cells and interstitial fluid. Interstitial fluid is organic (collagen type I) and inorganic (the most common are calcium and phosphate ions, along with bicarbonate, magnesium, potassium and sodium ions). Calcium and phosphate ions form hydroxyapatite crystals, from which the panels located along the collagen fibers are made. The bond between collagen fibers and crystals provides strength and resilience of the bone. While the bone keeps its shape when decalcified, it also becomes soft and pliable. If we remove the organic interstitial fluid, the bone's shape is maintained, but it becomes fragile and brittle (Petrovič and Zorc 2005). Kemp and Smith (2005) advocate a hypothesis that states that the binding of DNA to hydroxyapatite (which represents the main bone mass) provides stability of DNA and its preservation in old skeletal remains. The mechanism of DNA binding to hydroxyapatite is precarious, and it is also assumed that negatively charged phosphate groups of DNA molecules bind to hydroxyl groups on hydroxyapatite. This hypothesis is confirmed by the fact that when the degradation of hydroxyapatite increases, much less DNA can be preserved in the bone. Strong binding of collagen and non-collagen proteins on hydroxyapatite prevents its degradation, which might otherwise occur due to external influences (temperature and various chemical agents).

Edson et al. (2004) stated that the most suitable bones for DNA tests on skeletal remains are the long bones, especially the femur and rib bones, followed by tibiae and vertebrae, pelvic bones, humeri, blades and jaws with teeth; the skull bones are the least suitable for genetic investigation according to Edson and his colleagues (2004). From the comparative study of the performance of nuclear DNA typing of skeletal remains (we typed teeth, femurs and tibiae) from the mass graves of the Second World War, our laboratory discovered that teeth are the most suitable for typing, followed by the femur bones and tibiae (Zupanič Pajnič et al. 2012a; Zupanc et al. 2013). Similar conclusions were also reached by Miloš et al. (2007) and Misner and colleagues (2009).

The first successful DNA investigations carried out on bones that were between a few months and 10 years old were performed by Lee et al. (1991), Hochmeister et al. (1991, 1995) and Hagelberg et al. (1989, 1991). They were followed by numerous researchers who, depending on the age and degree of porosity of the bones, analysed nuclear microsatellites (Yamomoto et al. 1998; Tahir et al. 2000; Lleonart et al. 2000; Alonso et al. 2001; Miloš et al. 2007; Davoren et al. 2007; Irwin et al. 2007a; Irwin et al. 2007b; Zupanič Pajnič 2008; Vanek et al. 2009; Bogdanowicz et al. 2009; Zupanič Pajnič et al. 2010; Lee et al. 2010) or mtDNA polymorphisms (Sullivan et al. 1992; Fisher et al. 1993; Lutz et al. 1996; Jehaes 1998; Bender et al. 2000; Sco et al. 2000; Anslinger et al. 2001; Stone et al. 2001; Koyama et al. 2002; Palo et al. 2007). Hagelberg et al. (1989) performed molecular anthropological investigations on bones that were from 300 to 5500 years old and concluded that the preservation of DNA in them depends highly on the environment where the remains were located rather than their actual age. Lee et al. (1991) obtained from 0 to 500 ng of DNA per mg of bone out of a specimen that was a couple of months old and did not undergo previous decalcification; Hochmeister et al. (1991) managed to obtain from 25 ng to 3.3 µg of DNA/g from bones between a few months and 11 years old with previous decalcification; Yamamoto et al. (1998) obtained 5 ng of DNA from 1 gram of femur of a 16 year old skeleton of a new-born child; Seo et al. (2000) obtained 6 µg of DNA from 1.5 g of bone belonging to a 10 year old skeleton of a new-born child; Tahir et al. (2000) obtained 8 ng of DNA per gram of bone from a 27 year old exhumed skeleton. In our laboratory, we managed to obtain up to 100 ng of DNA per gram of bone powder and also make complete nuclear genetic profiles out of almost 70 year old bones found in the Second World War karst cave Konfin I mass grave (Zupanič Pajnič et al. 2010). We also extracted up to 10.7 ng DNA/g of tooth powder from the Auersperg chapel archaeological site skeletal remains and obtained a complete genetic profile of autosomal DNA, Y-STR haplotype and mtDNA haplotype for HVI and HVII region from one skeleton (Zupanič Pajnič et al. 2012b; Zupanič Pajnič 2013a)

At the Laboratory of Molecular Genetics of the Institute of Forensic Medicine, we select for genetic testing one long bone (preferably femur) and two teeth (preferably well-preserved and endodontically untreated molars) from each individual skeleton found in the Second World War graves; that is only possible through the excavation of skeletons in anatomic position.

In Figure 8 there are photos of selected bone and tooth samples for molecular genetic analyses.

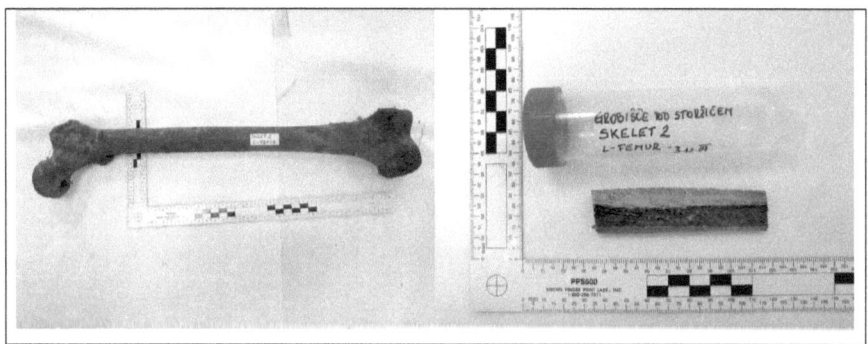

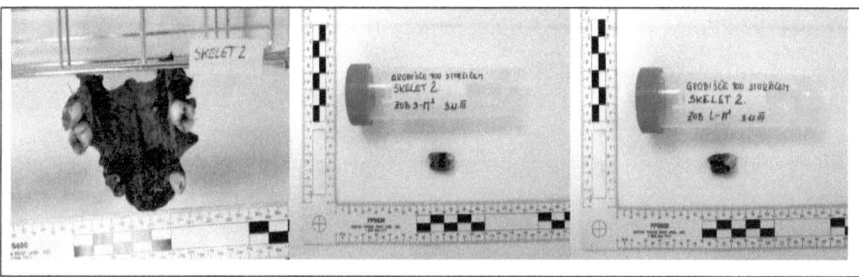

Figure 8: Photos of selected bone and tooth samples for molecular genetic analyses (photo: Rudi Bevc).

If excavation of skeletons was not carried out in the anatomical position, we select for molecular genetic investigations all left or all right femurs found in the grave.

In Figure 9 there are photos of sampled right femurs from the Second World War mass grave Konfin I.

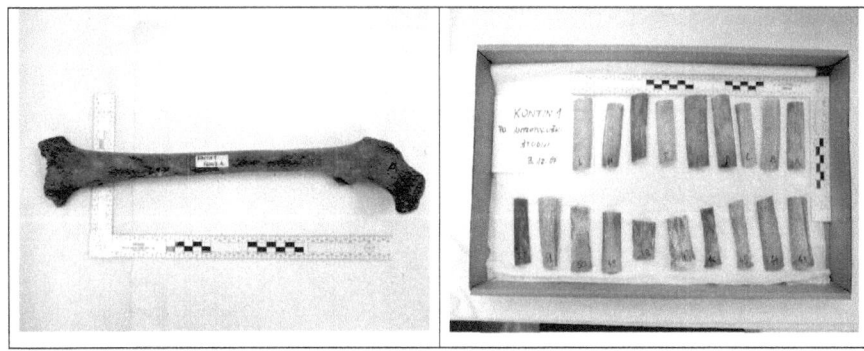

Figure 9: Photos of sampled right femurs from the Second World War mass grave Konfin I (photo: Rudi Bevc).

All skeletal material is photo-documented, appropriately labelled and frozen at -20°C until the DNA isolation procedure.

Typing of nuclear DNA and mtDNA is carried out for the bones, reference persons and persons to be included in the elimination database. For reference persons, mtDNA typing is carried out for maternal relatives and Y-chromosome typing for paternal relatives (Zupanič Pajnič 2013c).

MEASURES FOR PREVENTING DNA CONTAMINATION

Many chemical and physical environmental factors can have influence on contamination of human body remains with DNA from bacteria and fungi. That kind of contamination is not possible to prevent. It can affect the success of mitochondrial DNA and nuclear DNA typing. On the other hand, we are able to impact the prevention of the most important contamination-with modern DNA. Old and degraded samples possess a very low quantity of DNA and are therefore very susceptible to contamination with modern DNA. During PCR reaction, modern DNA competes with endogenous DNA from bones and teeth, so the worst outcome from such contamination is a genetic profile belonging to modern DNA. Contamination with modern DNA can occur during the excavation and storing of body remains. It can also occur in the molecular laboratory during the process of DNA typing. Surface contamination can often occur due to improper handling of skeletal remains with bare hands. We eliminate surface contamination through different methods. The most important are: washing in bi-distilled water, detergent, acids, ethanol and bleach; radiation with UV light and removing the bone surface with brush; and acquiring the bone or tooth material directly from the inside of the specimen. For successful decontamination, we usually use a combination of all listed methods.

One of the important studies on the use of bleach (NaOCl-sodium hypochlorite) for decontamination from the surface of bones and teeth was done by Kemp and Smith (2005). They discovered that bleach destroys the contaminating DNA on the surface while washing the material with 3-5% bleach for 15 minutes before DNA isolation. NaOCl, with its oxidative activity, splits contaminating DNA into short fragments or even single bases. That fact was successfully proven by the work of two researchers - Prince and Andrus (1992). They did not succeed in amplifying the 76 bp long DNA fragment previously soaked with 10% NaOCl (sodium hypochlorite; Clorox). Bleach does not harm endogenous DNA. It stays undamaged even if one uses 6% bleach for 21 hours; research done by Kemp and Smith (2005). The reason for the great stability of DNA in old skeletal remains is its linkage with hydroxyapatite, which protects the endogenous DNA from chemical degradation.

According to Kemp and Smith (2005), only physical removing of surface contamination is not sufficient because contaminating DNA is able to penetrate deeper into the bone. Therefore, it is necessary to use both of these techniques and to combine them with other methods. For successful surface decontamination, Kemp and Smith (2005) recommend washing in 6% bleach for 15 minutes. To remove the bleach, he washed the specimen for 1-2 minutes with ultra-pure bi-distilled water several times. With this kind of a technique, he successfully removed the surface contamination from 500 to 10,000 year old skeletal remains. Rennick et al. (2005) studied the influence of different techniques for removing the soft material from the surface of bones on degradation of endogenous DNA. He discovered that 4 hour cooking in 3% bleach caused degradation of endogenous DNA. Alternatively, he found that cooking in water or a water solution of detergent did not degrade DNA. He therefore recommends using detergent Alconox instead of bleach for removing the soft tissues from the surface of the bones.

Tooth pulp is the main source of DNA in the teeth. It is protected by enamel, dentin and dental cement and is less prone to contamination (Tsuchimochi et al. 2002). Most of the researchers (Sweet and Hildebrand 1998; Baker et al. 2001; Gaytmen and Sweet 2003) for decontamination use washing in detergent, ethanol and water in succession; the final step is UV illumination. In the case of a damaged tooth, surface decontamination is done only by washing out in distilled water and UV light illuminating.

Another type of contamination (besides surface contamination with modern DNA) is also possible - this is contamination with PCR products and contamination of reagents, laboratory plastic and other equipment. At the Laboratory of Molecular Genetics in the Institute of Forensic Medicine, we follow international recommendations for prevention of contamination (Pääbo 1989; Pääbo 1990; Wilson et al. 1995; Bär et al. 2000; Carracedo et al. 2000; Kalmar et al. 2000; Tully et al. 2001; Alonso et al. 2001; Pääbo et al. 2004;

Kemp and Smith 2005; Tamariz et al. 2006; Davoren et al. 2007; Shaw et al. 2008). We use the following measurements to prevent contamination in the laboratory:

1. To prevent contamination with own biological material, we always use clean, sterile gloves (we use double laboratory gloves) that we frequently change, surgical masks to cover the nose and mouth, caps and clean laboratory coats. We change gloves for every new sample
2. We clean the entire surface before and after the work with commercial disinfectant, water and ethanol. We use paper towels that we throw away after use. We clean the surface even between working with different skeletal remains
3. We use water and laboratory plastics that are definitively DNA free (according to manufacturer's guarantee)
4. We clean all tools for cleaning, abrasion and grinding of bones and teeth before use with 10% detergent *Alconox,* sterile bi-distilled water and 80% ethanol and sterilize them
5. We put all the reagents, tools and laboratory plastics after sterilization under UV light (for whole night). We expose all the listed material to the UV light for 30 minutes directly before starting to work.
6. We take new and clean tools for each bone or tooth specimen
7. We take new and clean gloves for each bone or tooth specimen, and we change the gloves even between treatment of that particular specimen if needed
8. We do not touch our face with the gloves
9. We analyse bone and teeth samples separately from the reference samples and we also separate them from the samples for elimination database (from the isolation of DNA to the DNA typing)
10. To prevent contamination with previously amplified products, we have different rooms (places) to separate each step in the bone typing

procedure. We have rooms for cleaning and grinding the bones and teeth, DNA isolation, PCR mixture preparation and analysis of amplified products. Amplified products from the analytical room never return to rooms for cleaning, isolation of DNA and PCR mixture preparation. To summarize, separation of pre and post PCR procedures must be provided.

11. We store all the reagents according to the manufacturer's recommendation
12. To detect any possible contamination with DNA or previously amplified PCR products of reagents or laboratory plastics, we always use negative control in the PCR
13. For monitoring the cleanliness of the isolation reagents and laboratory plastics, we always use isolation negative control
14. While sequencing the mtDNA, we sequence the negative control along with the samples
15. We always use filter tips to prevent aerosol contamination. They are exposed to UV light approximately 30 minutes before use at least
16. We use the room for cleaning, grinding and isolating the DNA from bones and teeth exclusively for this kind of biological material and not for any other sample that contains much more DNA (saliva, blood samples)
17. We isolate DNA from bones and teeth two times to check the results of genotyping
18. We sequence mtDNA in both directions to provide correctness of nucleotide sequence

EXCAVATION AND STORAGE OF SKELETAL REMAINS

Old skeletal remains contain very little DNA that can be highly degraded. Therefore, proper measures for excavating, anthropological examination and storing should be considered in order to succeed in molecular genetic identification. Incorrect handling and storing of skeletal remains can lead to contamination or degradation of endogenous DNA of bones and teeth and give the wrong results. Molecular genetic analysis can even fail due to this reason (Kolman and Tuross 2000).

One should consider the following facts when facing the excavation of skeletal remain. The use of clean gloves and facial masks (especially in the cases of illness like coughing and sneezing) is mandatory. Excavation according to anatomic position is recommended whenever possible.

In Figure 10 there are photos of the excavation of Second World War victims according to anatomic position.

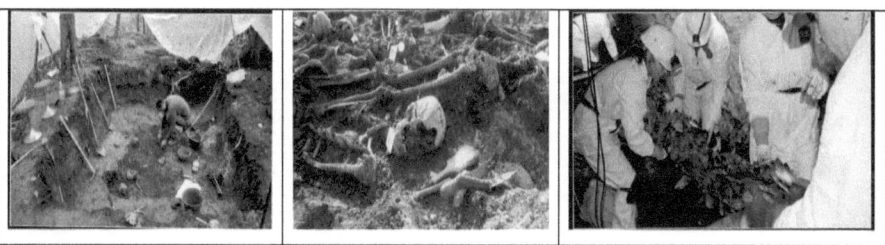

Figure 10: Photos of excavation of Second World War victims according to anatomic position (photo: Pavel Jamnik - left and middle; Jože Jagrič - right).

Skeletal remains should be stored in aerial boxes. Plastic bags are not suitable, because bones can't dry in them and the process of decay can start.

In Figure 11 there are photos of numbered skeletons (left), each of which is stored in a plastic aerial box (right).

Figure 11: Photos of numbered skeletons (left), each of which is stored in a plastic aerial box (right) (photo: Policijska uprava Maribor).

Different parts of the body (skull with teeth) should be stored in separate paper bags to protect them from falling apart. All fragments should be marked to ensure the belonging of each part to a particular skeleton. Regular paper is the most useful for such storing purposes. All skeletal remains should be photo documented. Boxes with marked skeletal remains should be stored in dry places with low humidity to minimize the possibility for development of microorganisms.

Before anthropological examination of the excavated remains, the laboratory should provide the anthropologist with all the necessary material for preventing contamination (sterilizing liquid, gloves, caps, masks and coats for single use only). The anthropologist should also be informed of how to handle remains in order to prevent contamination (the use of protective coats, changing gloves and sterilizing the working surfaces).

ELIMINATION DATABASE

As mentioned above, contamination of human biological remains can occur during excavation, anthropological examination, molecular genetic analysis and storage of the remains (Brown and Brown 1992). Within the genetic laboratory, contaminating DNA can be located on the laboratory plastics and reagents. DNA fragments can also be present in the air bind on the aerosol particles (Graham 2007). To be able to trace down any possible contamination, it is necessary to create an elimination database where all the profiles of persons that come in contact with samples are stored. An elimination database should contain profiles of persons who excavated, stored and examined (anthropologists, molecular geneticists) the remains during any stage of the process. Therefore, it is necessary to obtain buccal swabs of all participating persons. The laboratory should provide them with material for easy taking of buccal swabs (sterile cotton stick) and thorough instruction on how to take them. Furthermore, all the participants have to sign an informed consent to confirm that they agree with the fact they will be genotyped. For a complete elimination database, nuclear, mitochondrial and Y chromosome (for men only) genetic profiles should be ensured (Zupanič Pajnič 2008; Zupanič Pajnič et al. 2010). To be able to trace the contamination, it is also necessary to include extraction negative controls in every batch of extraction and PCR-negative controls in every amplification reaction to verify the purity of the extraction and amplification reagents and plastics.

FAMILY REFERENCES - RELATIVES OF THE VICTIMS

To inform living relatives (reference persons) of the victims how to cooperate with the investigating laboratory, it is convenient to publish the list of potential victims of the mass graves in local media together with the telephone number and the address of the laboratory that conducts the investigation. The laboratory staff should give any further answers to the questions relatives might have regarding buccal swab taking. After complete explanation and choosing of the right reference persons, all participating relatives must be sent a kit for self-taking buccal swabs together with instructions. The complete kit contains: sterile cotton stick, clean gloves and instructions. There should be an informed consent form included in the shipment. With their signing, the participants confirm to agree with the genetic identification investigation of the skeletal remains and use of their profiles as references. The relatives return the swabs and their written consent by post. For identification purposes, we choose the following reference persons:

- to compare autosomal microsatellites of nuclear DNA, we choose the closest relatives as reference persons;
- to compare polymorphisms of mitochondrial DNA, we choose relatives on the mother's side that could be in close or distant connection with the victim;
- to compare haplotypes of Y chromosome, we choose close or distant relatives on the father's side.

We have to gain genetic profiles of nuclear DNA for all the participating relatives, Y chromosome profiles for relatives in the father's family line and mtDNA profiles for the mother's family line.

When brothers are used as the family reference, the maternal and paternal line can be traced with mtDNA and Y-STRs along with autosomal STRs for comparison of close relatives. When sisters are used as the family reference along with autosomal STRs for comparison of close relatives, only the maternal line can be traced with mtDNA. Only autosomal STRs can be used for comparison when daughters are used as the family reference. In the latter case, distinct relatives are needed for comparison of Y-STRs (paternal line) and mtDNA (maternal line).

Molecular genetic identification of war mass grave victims has more steps. The first one is gaining DNA from skeletal remains, saliva of the relatives and persons performing the investigation. The next is DNA quantification of the isolated DNA and finally genotyping of nuclear, mitochondrial and Y chromosome DNA of bones and teeth as well as reference and elimination base persons. The cleanliness of the extraction blind control and amplifying negative control must be checked. Bones and teeth genetic profiles should be compared with genetic profiles of elimination database persons and afterwards with the profiles of the relatives. The final step is statistical evaluation of family relationship for matched genetic profiles of bones and teeth of victims with reference persons.

EXTRACTION OF GENOMIC DNA

The method for isolating DNA from bones and teeth is one of the most time consuming activities in the field of forensic genetics. A lot of inhibitory substances that are present and low DNA content are the reason to be very careful which method one chooses for isolating. Since methods used for DNA extraction from old skeletal remains also have a strong effect on amplification success, it is important to use an efficient extraction procedure. The method should be able to remove as many inhibitors as possible and to gain the maximum available DNA (Cattaneo et a. 1997). Decalcification with 0.5 M EDTA enables separation of bone cells from the bone mass (Bender et al. 2000). When working with fresh bones and teeth, decalcification is not needed. This step is very important for old skeletal remains, because decalcification is crucial for gaining higher quantities of DNA (Hochmeister et al. 1991). Loreill and co-workers (2007) succeeded in gaining a sufficient quantity of DNA when they used complete demineralization from old skeletal remain that gave no results without it. As shown by the latest studies, total demineralization is the best method of DNA extraction from old bone material (Jakubowska et al. 2012; Amory et al. 2012), since total demineralization significantly increases the proportion of full profiles, reflecting a correlation with better DNA quality.

Results from many research works have shown that washing and improper handling of remains contaminate their surface and can even penetrate into deeper layers. This kind of inner contamination is dependent on the stage of porosity and preservation of the remains (Salmon et al. 2005; Gilbert et al. 2005; Sampierto et al. 2006). Therefore, skeletal remains must be cleaned mechanically and chemically and be UV irradiated.

The method of DNA extraction was developed in our laboratory to acquire high quality DNA from Second World War skeletal remains and from skeletal remains from archaeological sites. The same method is also used in our laboratory for molecular genetic identification of unknown decomposed bodies in routine forensic casework where only bones and teeth are suitable for DNA typing. We analysed 111 bones and teeth from Second World War mass graves to evaluate this method (Zupanič Pajnič 2011a) and additionally 54 Second World War samples and some bones and teeth from archaeological sites for change extraction protocol from partial to total demineralisation (Zupanc et al. 2013; Zupanič Pajnič 2013a).

We cleaned the bones and teeth, removed surface contaminants and ground the bones into powder using liquid nitrogen. Prior to isolating the DNA using the

BioRobot EZ1 (Qiagen), the powder is decalcified. The whole procedure is carried out in a room designed exclusively for processing old skeletal remains. Mechanical cleaning is performed in a closed cytostatic safety cabinet.

In Figure 12 there are photos of cutting, mechanical cleaning, grinding and extracting of DNA from Second World War bones.

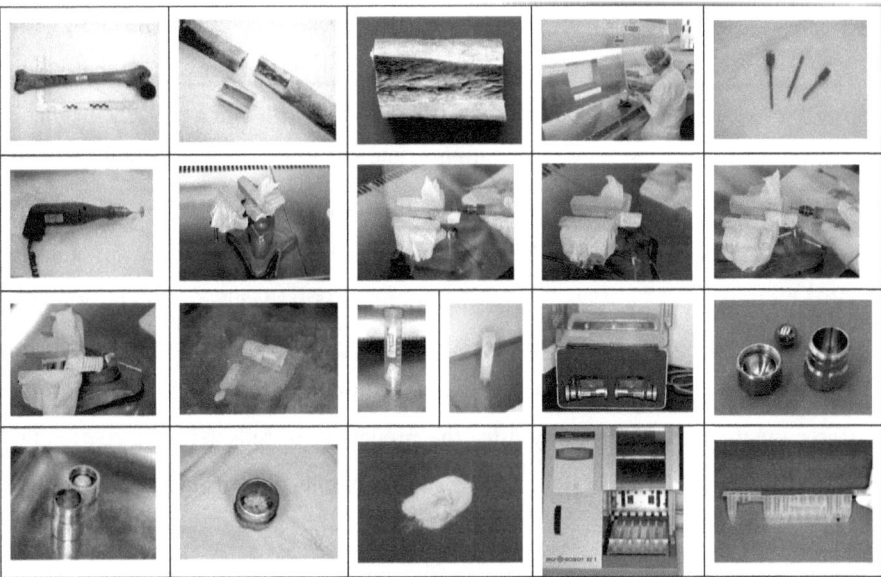

Figure 12: Photos of cutting, mechanical cleaning, grinding and extracting of DNA from Second World War bones (photo: Rudi Bevc).

The bone samples are cleaned mechanically and chemically. The surface is decontaminated by physical removal of the surface using a rotary sanding tool (Dremel) and rinsing in 5% Alconox detergent, water and 80% ethanol. Grinding in a TissueLyser (Retsch) homogenizer using liquid nitrogen followed. Genomic DNA is obtained from 0.5 g of bone or tooth powder incubated in 8.5

ml of 0.5 M EDTA pH 8.3 for 24 h at 37°C for decalcification. After centrifugation, the supernatant is discarded, and the precipitate is washed with ultrapure distilled water, which is discarded after centrifugation. An extraction buffer, proteinase K and DTT are added to the precipitate and incubated overnight at 56°C. After centrifugation, the supernatant is taken to isolate DNA. The DNA is purified in a Biorobot EZ1 (Qiagen) device using the EZ1 DNA Investigator Card and EZ1 DNA Investigator Kit (Qiagen). The whole procedure is automated and takes about 20 minutes to complete. It doesn't use any aggressive organic solvents like phenol or chloroform. It is based on technology of magnetic particles that are covered with silicon. Such magnetic particles are very efficient for binding DNA, especially in the presence of chaotropical salts (like guanidine thyocyanat GuSCN or guanidine hydroksychloride-GuHCl or NaJ). These salts are highly efficient for nucleic acid cleaning (Boom et al. 1990). Chaotropic salts lyse cells, denature proteins, inactivate nucleases and accelerate the binding of DNA to the paramagnetic particles covered with silicon. The whole extraction process is done in a huge filter tip that is thrown away after the procedure is finished. The rest of the extraction reagents are safely placed in a container - cartridge for single use only. Therefore, no manual pipetting is needed. This is very important for prevention of contamination. High efficiency of magnetic particles in DNA extraction was confirmed in several studies (Nagy et al. 2005; Montpetit et al. 2005; Kishore et al. 2006; Valgren et al. 2008). We freeze the DNA that was extracted until forthcoming steps of quantification and genotyping of nuclear and mtDNA. We always include negative controls in the process of extraction to check cleanliness of laboratory plastics and reagents.

We isolate the DNA of reference and elimination database persons with the use of a EZ 1 DNA Investigator Kit, the same named software card (Qiagen) and the BioRobot EZ1 instrument (Qiagen).

DNA QUANTIFICATION

When we work with highly degraded DNA or with samples with low quantity of DNA, we use DNA quantification to ensure the quality of assays based on the PCR. Because of the stochastic effect at amplification of the samples with very low quantity of DNA (less than 20 copies - 60 pg for nuclear DNA), the possibility of errors is much higher due to unamplified alleles (Alonso et al. 2004). There are many quantification methods used for measurement of DNA quantity in samples. The newest is based on real time quantification that was found to be the most sensitive so far (Alonso et al. 2003; Alonso et al. 2004; Andreasson et al. 2006). Correct DNA quantification is an essential part to obtaining reliable STR typing results. Some important information about the compromised bone and tooth samples can be obtained using commercial real-time PCR quantification kits, the most often used for DNA quantification by forensic DNA analysts. We can use quantification as a screening tool to decide whether the sample is suitable for nuclear or mtDNA typing. We determine the optimal DNA quantity for successful genotyping. In this way, we can save the sample for further independent analysis (Noonan et al. 2006). DNA quantification also enables detection of inhibitory substances in the sample (Alonso et al. 2003).

Before carrying out our PCR, we quantitate every sample (from bones, teeth and saliva from relatives and elimination database persons) with the method of real - time DNA quantification. This method is based on 5' nuclease activity of Taq DNA – polymerase, and it uses fluorescent dyes with marked probes combined with the instrument's detection system in which promotion of PCR amplification is monitored in real time. The software automatically calculates the DNA concentration of a sample with the use of a standard curve. Amplification of internal positive control (IPC) that is a part of PCR mixture and therefore present at each sample is an indicator of possible inhibitors. Incorrect DNA quantification due to the presence of PCR inhibitors may affect experiment results and the CT values of an internal PCR control must be carefully checked (Seo et al. 2012).

After DNA quantification, we succeeded in obtaining up to 100 ng DNA/g of bone powder from 82 femurs excavated from the Second World War mass grave in the karst cave Konfin I (Zupanič Pajnič et al. 2010); up to 8.6 ng DNA/g of bone powder from three femurs from the grave under Mount Storžič (Zupanič Pajnič 2008) and up to 16 ng DNA/g of bone powder from 25 femurs excavated from the mass grave at Bodovlje Gorge (Bodoveljska grapa) (Zupanič Pajnič 2007). We also extracted up to 10.7 ng DNA/g of tooth powder from the Auersperg chapel archaeological site skeletal remains (Zupanič Pajnič et al. 2012b; Zupanič Pajnič 2013a). We didn't detect any nuclear DNA in any isolation blind controls, which shows the cleanliness of the isolation procedure and also that no contamination of samples during the isolation procedure occurred.

According to the different length of DNA fragments amplified using different quantification kits, the degree of degradation in compromised samples can be estimated. In order to estimate the degree of DNA degradation in old bone and tooth samples, we evaluated two commercially available kits on 54 Second World War bones and teeth (Zupanc et al. 2013). The Quantifiler Human DNA Quantification Kit (Applied Biosystems) and recently released the Investigator Quantiplex kit (Qiagen) were used for quantification. The Quantifiler amplifies 62 base pairs (bp) amplicon and the Quantiplex kit amplifies 146 bp amplicon of nuclear DNA. Using the commercially available Human Quantifiler kit and Quantiplex kit, it is possible to determine the quantity of DNA in the range from 20 pg DNA/µl up to 50 ng DNA/µl of sample. The quantities determined with amplification of 62 bp fragment (Quantifiler kit) were up to 6 times higher in almost two thirds of the samples than the quantities determined with amplification of 146 bp fragment (Quantiplex kit), implicating degradation of nuclear DNA. Degradation was observed in 90% of tibias, 80% of femurs and only 30% of teeth, indicating DNA in bones is more degraded than DNA in teeth from the Slovenian Second World War victims analysed. These findings are in accordance with our previous study using different amplification kits where more complete autosomal STR profiles were obtained from old teeth than from old bones (Zupanič Pajnič et al. 2012a).

DNA TYPING

For molecular genetic analyses of Second World War bones and teeth, autosomal STRs have to be typed and Y-chromosomal and mtDNA haplotypes have to be determined. For autosomal STR typing, three commercially available amplification kits are used in our laboratory: the NGM PCR Amplification Kit (Applied Biosystems), the Investigator ESSplex plus Kit (Qiagen) and the MiniFiler PCR amplification kit (Applied Biosystems). Genetic profiles acquired with the NGM Kit are verified with the Investigator ESSplex plus Kit and when degraded samples are examined also with the MiniFiler kit. For Y-STR typing, we use the AmpFlSTR Yfiler PCR Amplification Kit (Applied Biosystems) and we repeat the amplifications at least twice. For determining the mtDNA haplotypes, we sequence two hypervariable regions of mtDNA control region HVI and HVII in both orientations in order to verify the accuracy of base-calling. In cases of length heteroplasmy in the poly-C strand, the polymorphisms behind the C-stretch in the forward sequencing reaction and before the C-stretch in the reverse reaction are confirmed by repeating amplification and sequencing reactions according to the recommendations

(Tully et al. 2001). Following Bandelt and Parson (2008), in cases of heteroplasmic length variants the dominant variants are reported.

In reference persons, typing of autosomal DNA has to be performed. We use the NGM Kit or the Investigator ESSplex plus Kit. When statistical analyses do not reveal a posterior probability of 99.9%, reference samples are typed additionally with the MiniFiler kit and with the PowerPlex ESX 17 System to obtain the profiles on loci D13S317, D7S820, CSF1PO and SE33. For the maternal relatives, mtDNA haplotypes also have to be obtained and for the paternal relatives, Y-STR haplotypes using the Yfiler Kit.

In persons from the elimination database, in addition to autosomal DNA typing using the NGM Kit or the Investigator ESSplex plus Kit, typing of mtDNA has to be performed and for males also typing of Y-STRs using the Yfiler Kit.

AUTOSOMAL STR DNA TYPING

The chemistry and methods used for DNA amplification may have a strong effect on the amplification success when dealing with compromised old skeletal DNA samples (Putkonen et al. 2010; Irwin et al. 2012). In 2009 and 2010, new amplification kits were developed to meet the European Network of Forensic Institutes and the European DNA Profiling group recommendations for increasing the European Standard Set (ESS) of loci to improve its discrimination power and to fulfil the increasing requirements regarding sensitivity and reproducibility for the analysis of minute amounts of DNA by adopting five additional mini-STRs: D2S441, D10S1248, D22S1045, D1S1656 and D12S391 (Gill et al. 2006a; Gill et al. 2006b). Some validation, concordance and population studies (Parys-Proszek et al. 2010; Poetsch et al. 2010; Budowle et al. 2010; Gill et al. 2011; Yurrebaso et al. 2011; Molnar et al. 2011; Tucker et al.

2011; Albinsson et al. 2011; Hill et al. 2011; Previdere et al. 2011; Kutranov 2011; Scherer et al. 2011; Garcia et al. 2012; Welch et al. 2012) have been published for new amplification kits with the extended ESS of loci. It was shown that the new kits are robust enough to genotype degraded DNA samples through the use of mini STR loci and have increased tolerance to common inhibitors and increased sensitivity to obtain full profiles from low-level DNA samples from casework (Tucker et al. 2011; Sprecher et al. 2009; Poetsch et al. 2011). We evaluated the performance of different commercially available amplification kits on old skeletal remains and found out that the ESX 17 (Promega), the Investigator ESSplex and the Investigator ESSplex plus (Qiagen), the NGM (Applied Biosystems), the Identifiler (Applied Biosystems) and the PowerPlex 16 (Promega) kits are all highly reliable for STR typing of Second World War skeletal remains with the DNA extraction method optimised in our laboratory (Zupanič Pajnič et al. 2012a; Zupanc et al. 2013; Zupanič Pajnič 2013b). We found out that the first four amplification kits can be used for typing of Second World War bones and teeth following the PCR protocols recommended by the manufacturers without increasing the number of cycles or any other modification of protocols, while for the latter two older kits, the amplification protocol has to be changed for low amount DNA samples. In addition to increasing the number of cycles and extending the elongation time, BSA has to be added, double amount of AmpliTaq Gold DNA Polymerase has to be used, the amount of Primer Pair Mix has to be increased and the final extension step has to be prolonged. The number of cycles has to be increased only for samples with a low amount of DNA by 3 in the Identifiler and 2 in the PowerPlex 16 amplification kit (Zupanič Pajnič et al. 2010; Zupanič Pajnič 2013b).

For autosomal STR typing of Second World War skeletal remains, we routinely used the Identifiler PCR Amplification Kit (Applied Biosystems), the PowerPlex

16 System (Promega) and additionally, for degraded DNA samples, the MiniFiler PCR Amplification Kit (Applied Biosystems). The Identifiler Kit and the PowerPlex 16 System contain the same 13 core STR loci and amelogenin, whereas the Identifiler Kit also contains the loci D2S1338 and D19S433, and the PowerPlex 16 System the loci Penta E and Penta D. Overall, 17 STR loci and amelogenin were amplified in the past.

In last three years, we routinely used, for autosomal STR typing of old bones and teeth, two new commercially available amplification kits with ESS loci. Genetic profiles acquired with the NGM PCR Amplification Kit (Applied Biosystems) are verified with the Investigator ESSplex plus Kit (Qiagen) and when degraded samples are examined also with the MiniFiler amplification kit (Applied Biosystems), which enables amplification of shorter fragments. The NGM Kit and the Investigator ESSplex plus Kit contain the same 15 core STR loci and amelogenin. The MiniFiler PCR Amplification Kit (Applied Biosystems) contains eight STR loci, of which five are shared with the NGM Kit and the Investigator ESSplex plus Kit, but it uses shorter amplicons, which makes them more likely to be successful on fragmented DNA. Overall, 18 STR loci and amelogenin are amplified. When statistical analyses do not reveal a posterior probability of 99.9%, the PowerPlex ESX 17 System (Promega) amplification kit is used to obtain the profiles on additional SE33 STR locus.

In reference samples - relatives and in persons to be included in elimination database - autosomal STR typing is performed with the NGM PCR Amplification Kit or Investigator ESSplex plus Kit. When statistical analyses do not reveal a posterior probability of 99.9%, reference samples are typed additionally with the MiniFiler kit and with the PowerPlex ESX 17 System.

With the AmpF/STR NGM™ PCR Amplification Kit (Applied Biosystems), 15 autosomal STRs (D3S1358, TH01, D21S11, D18S51, D10S1248, D1S1656, D2S1338, D16S539, D22S1045, vWA, D8S1179, FGA, D2S441, D12S391 and D19S433) and amelogenin for gender determination are amplified simultaneously. Optimal amount of started genomic DNA for NGM kit is 0.5 ng. Extraction negative controls and negative PCR controls are amplified in parallel with samples from skeletal remains. In the extraction negative controls, the maximum volume of extracts are used for amplification.

With the Investigator ESSplex plus Kit (Qiagen), 15 autosomal STRs (D3S1358, TH01, D21S11, D18S51, D10S1248, D1S1656, D2S1338, D16S539, D22S1045, vWA, D8S1179, FGA, D2S441, D12S391 and D19S433) and amelogenin for gender determination are amplified simultaneously. The optimal amount of started genomic DNA for the Investigator ESSplex kit is 0.35 ng. Extraction negative controls and negative PCR controls are amplified in parallel with samples from skeletal remains. In the extraction negative controls, the maximum volume of extracts are used for amplification.

With the MiniFiler PCR Amplification Kit (Applied Biosystems), 8 autosomal STRs (D13S317, D7S820, D2S1338, D21S11, D16S539, D18S51, CSF1PO in FGA) and amelogenin for gender determination are amplified simultaneously. The optimal amount of started genomic DNA for MiniFiler kit is 0.5 to 0.75 ng. Extraction negative controls and negative PCR controls are amplified in parallel with samples from skeletal remains. In the extraction negative controls, the maximum volume of extracts are used for amplification.

The fluorescent-labelled products of the amplification kits are separated on an automatic ABI PRISM™ 3130 Genetic Analyzer (Applied Biosystems) using the 3130 Performance Optimized Polymer 4 (Applied Biosystems) and the

GeneScan-500 LIZ (Applied Biosystems) internal size standard with the NGM kit and DNA size standard 550 BTO (Qiagen) with the ESSplex kit. The genetic profiles are determined using the Data Collection v 3.0 and GeneMapper ID v 3.2 (Applied Biosystems) computer software with a 50 relative fluorescence units (RFU) peak amplitude threshold for all dyes.

Genetic profiles are shown in the form of electropherograms on which each peak suits to one allele of the STR locus. In the NGM kit (Applied Biosystems), four STR loci are labelled in blue (D10S1248, vWA, D16S539 and D2S1338), three STR loci and amelogenin in green (D8S1179, D21S11 and D18S51), four STR loci in black (D22S1045, D19S433, THO1 and FGA) and four STR loci are labelled in red (D2S441, D3S1358, D1S1656 and D12S391). In the Investigator ESSplex plus kit (Qiagen), four STR loci and amelogenin are labelled in blue (THO1, D3S1358, vWA and D21S11), five STR loci in green (D16S539, D1S1656, D19S433, D8S1179 and D2S1338), four STR loci in black (D10S1248, D22S1045, D12S391 and FGA) and two STR loci are labelled in red (D2S441 and D18S51). In the MiniFiler kit (Applied Biosystems), two STR loci are labelled in blue (D13S317 and D7S820), two STR loci and amelogenin in green (D2S1338 and D21S11), two STR loci in black (D16S539 and D18S51) and two STR loci in red (CSF1PO and FGA).

In Figure 13 are the electropherograms of autosomal STR profiles of Second World War bones and teeth, obtained with the Identifiler PCR Amplification Kit (Applied Biosystems), the PowerPlex 16 System (Promega), the PowerPlex ESX 17 System (Promega), the MiniFiler PCR Amplification Kit (Applied Biosystems), the AmpF/STR NGM^{TM} PCR Amplification Kit (Applied Biosystems) and the Investigator ESSplex plus Kit (Qiagen).

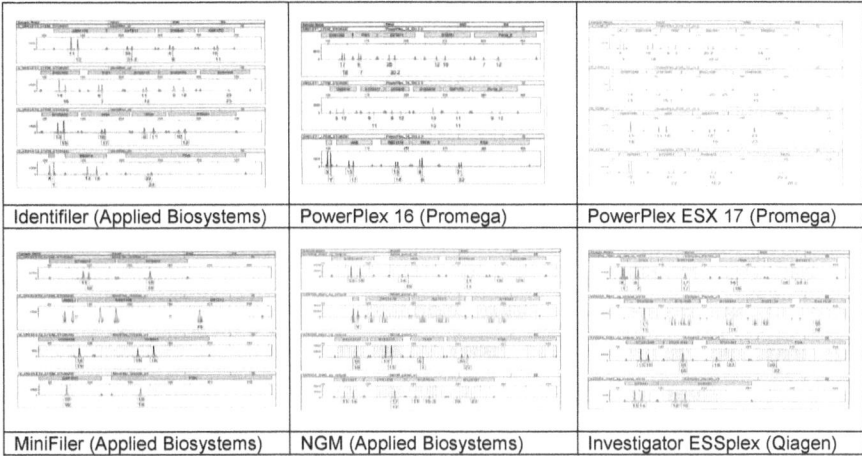

Figure 13: Electropherograms of autosomal STR profiles of Second World War bones and teeth, obtained with the Identifiler PCR Amplification Kit (Applied Biosystems), the PowerPlex 16 System (Promega), the PowerPlex ESX 17 System (Promega), the MiniFiler PCR Amplification Kit (Applied Biosystems), the AmpF/STR NGM^{TM} PCR Amplification Kit (Applied Biosystems) and the Investigator ESSplex plus Kit (Qiagen).

In molecular genetic identification of Slovenian Second World War victims, we managed to obtain autosomal STR genetic profiles in 98% of femurs from the Konfin I mass grave (Zupanič Pajnič et al. 2010), from all skeletons from the Storžič grave (Zupanič Pajnič 2008) and from 72% of femurs from the Bodovlje gorge (Bodoveljska grapa) mass grave (Zupanič Pajnič 2007).

Y CHROMOSOMAL STR DNA TYPING

With the AmpF/STR YFiler PCR Amplification Kit (Applied Biosystems), 17 STRs (DYS456, DYS389I, DYS390, DYS389II, DYS458, DYS19, DYS385 a/b, DYS393, DYS391, DYS439, DYS635, DYS392, Y GATA H4, DYS437,

DYS438 and DYS448) are amplified simultaneously. The optimal amount of started genomic DNA for YFiler kit is 0.5 to 1 ng. For old bone and tooth samples containing >50 pg/μl DNA, in our laboratory PCR is performed according to the manufacturer's instructions, but BSA (Sigma; final concentration 40 ng/μl) is added, and 1.2 μl AmpliTaq Gold DNA Polymerase (Applied Biosystems) is used. For samples with a concentration <50 pg/μl, in addition to the measures described above, the number of cycles is increased from 30 to 33, and the extension step within cycles is prolonged to 2 min. The PCR reaction is performed with at most 13 μl DNA, and we repeat the amplifications at least twice. Simultaneously with the samples from skeletal remains, extraction negative controls and negative PCR controls are amplified. In the extraction negative controls, the maximum volume of extracts are used for amplification.

The fluorescent-labelled products of the amplification kit are separated on an automatic ABI PRISM^{TM} 3130 Genetic Analyzer (Applied Biosystems) using the 3130 Performance Optimized Polymer 4 (Applied Biosystems) and the GeneScan-500 LIZ (Applied Biosystems) internal size standard. The genetic profiles are determined using the Data Collection v 3.0 and GeneMapper ID v 3.2 (Applied Biosystems) computer software with a 50 relative fluorescence units (RFU) peak amplitude threshold for all dyes.

Genetic profiles are shown in the form of electropherograms on which each peak suits to one allele of the Y-STR locus. In the YFiler kit, four STR loci are labelled in blue (DYS456, DYS389I, DYS390 and DYS389II), four STR loci in green (DYS458, DYS19, DYS385a and DYS385b), five STR loci in black (DYS393, DYS391, DYS439, DYS635 and DYS392) and four STR loci are labelled in red (Y GATA H4, DYS437, DYS438 and DYS448).

In Figure 14 there is an electropherogram of the Y - chromosomal haplotype of the tooth from skeleton 2 from the Storžič mass grave, obtained with the AmpF/STR Yfiler kit (Applied Biosystems).

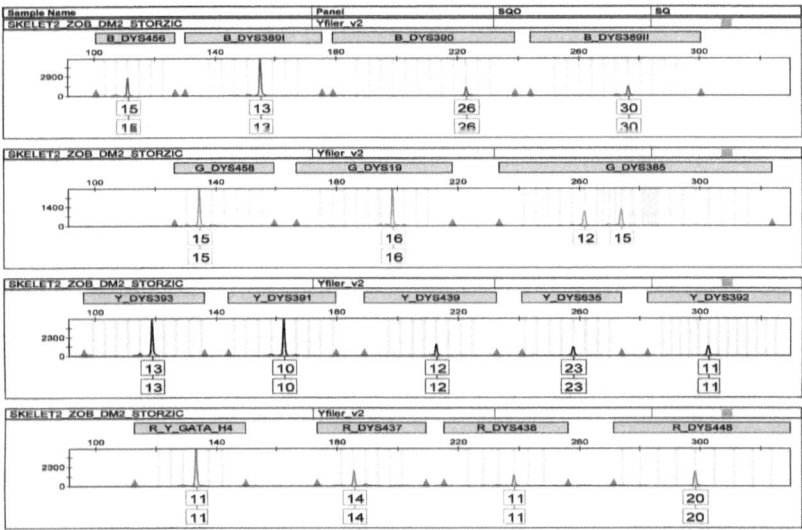

Figure 14: Electropherogram of Y - chromosomal haplotype of the tooth from skeleton 2 from the Storžič mass grave, obtained with the AmpF/STR Yfiler kit (Applied Biosystems).

In molecular genetic identification of Slovenian Second World War victims, we managed to obtain Y-chromosome haplotypes in 98% of femurs from the Konfin I mass grave (Zupanič Pajnič et al. 2010) and from all skeletons from the Storžič grave (Zupanič Pajnič 2008).

MITOCHONDRIAL DNA TYPING

For determining the mtDNA haplotype, the two hypervariable regions of mtDNA are amplified by the polymerase chain reaction (PCR) using the primers F15997/R16401 for HVI and F29/R408 for HVII (Parson et al. 1998). PCR is carried out according to Zupanič Pajnič et al. (2004). Prior to sequencing the PCR, products are purified using Centricon 100 spin dialysis columns (Amicon) following the manufacturer's recommendation. Sequencing reactions are performed in both orientations in order to verify the accuracy of base-calling. The primers used for the sequencing of the PCR products are the same as for the amplification. Sequencing reactions are carried out using the BigDye Terminator Cycle Sequencing Ready Reaction kit. The removal of excess dye-deoxy terminators, primers and buffers is accomplished with MicroSpin G-50 columns (Amersham Pharmacia Biotech) following the manufacturer's recommendations. Automated DNA sequencing is carried out on an ABI Prism 3130 Genetic Analyzer (Perkin Elmer) using the POP 4 (Applied Biosystems). Analysis of mitochondrial DNA sequencing data is performed using the Data Collection v 3.0 and the Sequencing Analysis Software, Version 5.2 (Applied Biosystems) computer software. The sequences are aligned and compared with the Anderson sequence (Anderson et al. 1981) from 16030 to 16401 for HVI region and from 55 to 408 for the HVII region.

Results of mtDNA sequencing are shown in the form of electropherograms on which each peak suits to one base (C is labelled in blue, A in green, G in black and T in red.

In Figure 15 there is an electropherogram of mtDNA sequence of the HVII region of the tooth from skeleton 2 from the Storžič mass grave, obtained with the BigDye Terminator Cycle Sequencing Ready Reaction kit (Applied Biosystems).

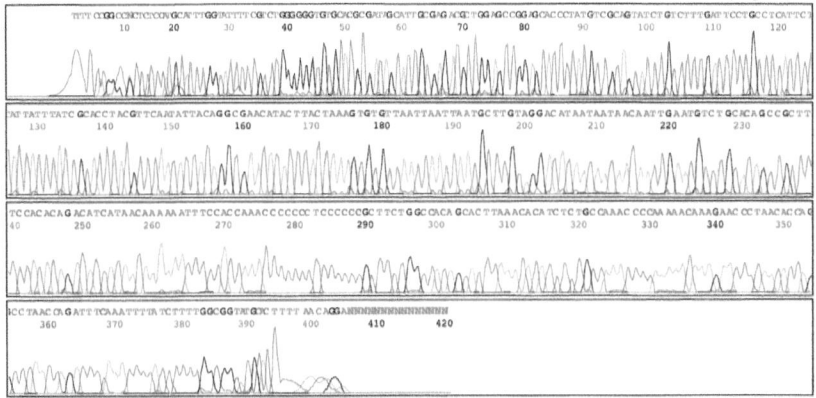

Figure 15: Electropherogram of mtDNA sequence of the HVII region of the tooth from skeleton 2 from the Storžič mass grave, obtained with the BigDye Terminator Cycle Sequencing Ready Reaction kit (Applied Biosystems).

In molecular genetic identification of Slovenian Second World War victims, we managed to obtain HVI and HVII mtDNA haplotypes in 98% of femurs from the Konfin I mass grave (Zupanič Pajnič et al. 2010), from all skeletons from the Storžič grave (Zupanič Pajnič 2008) and from all of the femurs from the Bodovlje gorge (Bodoveljska grapa) mass grave (Zupanič Pajnič 2007).

STATISTICAL ANALYSES OF FAMILIAR RELATIONSHIPS

Genetic profiles obtained from the bones and reference samples are compared, and estimation of potential familiar relationships is performed. The probabilities of relationship for autosomal STRs are calculated as a likelihood ratio (LR) and posterior probability (PP):

Posterior probability PP = LR x prior / (LR x prior + (1 − prior)) x 100%

Prior probability (prior) = 1/n+1 (n = number of victims in mass grave)

The calculation of likelihood ratios and posterior probabilities is performed with the DNAVIEW statistical software (Brenner 2007), using allele frequencies of the Slovenian population (Zupanič et al. 1998; Zupanič Pajnič et al. 2001) and prior probability of 1/n+1, where n represents the number of victims in the mass grave (Brenner and Weir 2007). Likelihood ratios for chromosome Y-haplotypes and mtDNA haplotypes are calculated based on haplotype frequencies in various databases (Gusmao et al. 2006). The counting method is used to estimate the haplotype frequencies, and the Balding and Nichols (1994) correction for errors in sampling is considered (Carracedo et al. 2000; Walsh et al. 2008). The reference database Y chromosome haplotype reference database - YHRD (Willuveit and Roewer 2007) and European mtDNA database - EMPOP (Parson and Dür 2007) are used to determine the Y-chromosome and mtDNA haplotype frequencies. In the YHRD database, we use European metapopulation, and in the EMPOP database, we use west Eurasian populations.

Whenever an agreement of autosomal genetic profiles and mtDNA haplotypes is noted between the bone and a relative, the product rule is used to estimate a combined likelihood ratio:

$LRc = LR_{(autosomal\ STRa)} \times LR_{(mtDNA)}$ (Castella et al. 2006).

The same is applied when an agreement is noted between autosomal genetic profiles and Y-STR haplotypes:

$LRc = LR_{\text{(autosomal STRs)}} \times LR_{\text{(Y-STRs)}}$ (Walsh et al. 2008).

Following recommendations (Biesecker et al. 2005; Brenner and Weir 2003; Prinz et al. 2007), the prior probability is set based on the number of mass grave victims reported, and a recommended posterior probability for kinship of 99.9% is used with the goal of high confidence of correct identification of victims in the mass grave. So, the statistical analyses show a high confidence of correct identification if the victim is identified with posterior probability higher than 99.9%.

EXAMPLES OF IDENTIFICATION OF THE SECOND WORLD WAR MASS GRAVE VICTIMS

The identification of Second World War mass grave victim is shown on an example of two Konfin I mass grave victims.

Victim 1 was identified through a comparison of the bone autosomal STR profile and the autosomal STR profile of the living daughter.

In Figure 16 we present right femur 41 and the fragment that was used for genetic analyses.

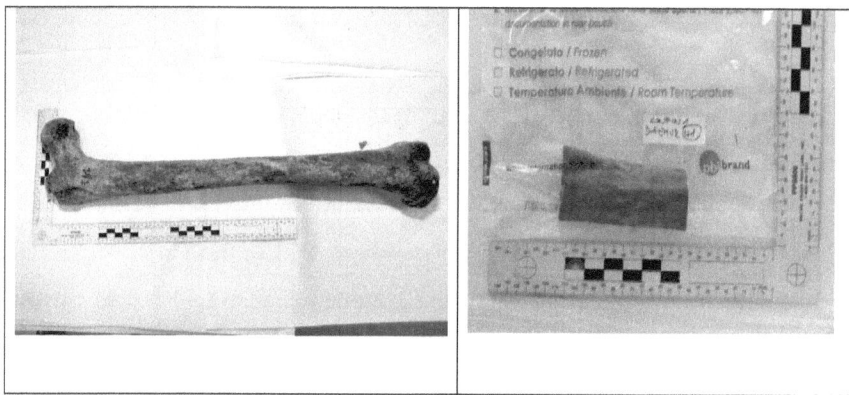

Figure 16: Right femur 41 from the Konfin I mass grave and the fragment that was used for genetic analyses (photo: Rudi Bevc).

In Table 1 we present the autosomal STR profiles from right femur 41 and the living daughter, obtained with the Identifiler kit (Applied Biosystems) and PowerPlex 16 amplification kit (Promega).

Table 1: Autosomal STR profiles from right femur 41 and the living daughter obtained with the Identifiler kit (Applied Biosystems) and PowerPlex 16 amplification kit (Promega).

Sample	D8S1179		D21S11		D7S820		CSF1PO		D3S1358		THO1		D13S317	
victim's daughter	8	13	29	30.2	8	12	11	11	15	16	6	9.3	11	11
R FEMUR 41	13	15	30.2	32.2	12	14	11	12	16	18	6	9	8	11
Sample	D2S1338		D19S433		vWA		TPOX		D18S51		Amelog.		D5S818	
victim's daughter	19	25	16	16.2	15	17	8	8	16	17	X	X	11	12
R FEMUR 41	20	25	15	16.2	15	18	8	10	14	17	X	Y	11	11
Sample	Penta E		Penta D		D16S539		FGA							
victim's daughter	5	9	9	14	12	12	19	23						
R FEMUR 41	9	18	9	13	12	12	23	23						

When comparing autosomal STR profiles of right femur 41 (putative father) and the victim's daughter, we include the paternity. Calculated paternity index (PI) was 5.9×10^5 and the posterior probability of paternity (PP) was 99.98%. The prior probability was set to 0.01 (number of victims in mass grave was 88).

Victim 2 was identified with the comparison of bone and the living sister autosomal STR profiles and mtDNA haplotypes.

In Figure 17 we present right femur 48 pat and the fragment that was used for genetic analyses.

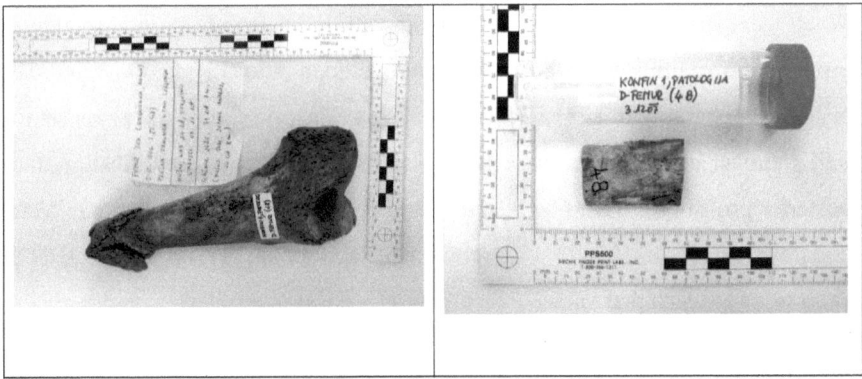

Figure 17: Right femur 48 pat from the Konfin I mass grave and the fragment that was used for genetic analyses (photo: Rudi Bevc).

In Table 2 we present the autosomal STR profiles from right femur 48 pat and the living sister, obtained with the Identifiler kit (Applied Biosystems).

Table 2: Autosomal STR profiles from right femur 48 pat and the living sister obtained with the Identifiler kit (Applied Biosystems).

Sample	D8S1179		D21S11		D7S820		CSF1PO		D3S1358		THO1	
R fem 48 pat	13	14	29	30	10	12	11	12	17	17	7	9.3
victim's sister	14	14	30	33.2	10	11	10	12	17	17	7	9.3
Sample	D2S1338		D19S433		vWA		TPOX		D18S51		Amelog.	
R fem 48 pat	20	25	12	15	16	16	8	8	12	18	X	Y
victim's sister	23	25	12	13	14	16	8	8	12	14	X	X
Sample	D13S317		D16S539		D5S818		FGA					
R fem 48 pat	8	13	13	13	11	11	22	23				
victim's sister	8	13	12	13	11	12	19	21				

When comparing autosomal STR profiles of right femur 48 pat (putative brother) and victim's sister, the calculated likelihood ratio (LR) was 681 and the posterior probability of sibling (PP) was 87.3%. The prior probability was set to 0.01 (number of victims in mass grave was 88). For positive identification, the posterior probability has to be 99.9% (Biesecker et al. 2005; Brenner and Weir 2003; Prinz et al. 2007). We combined autosomal STRs with mtDNA haplotypes to reach that value.

In Table 3 we present the mtDNA haplotypes from right femur 48 pat and the living sister.

Table 3: MtDNA haplotypes from right femur 48 pat and the living sister.

Sample	Differences according to "CRS"	mtDNA position
R fem 48 pat	HVI: 16093C, 16224C, 16311C	HVI: 16030–16381
	HVII: 73G, 152C, 263G, 315.1C	HVII: 55–388
victim's sister	HVI: 16093C, 16224C, 16311C	HVI: 16030–16381
	HVII: 73G, 152C, 263G, 315.1C	HVII: 55–388

MtDNA haplotypes of right femur 48 pat and the victim's sister are identical. The same haplotype was found only once in the EMPOP database (European Caucasians), and the calculated likelihood ratio (LR) was 1493. For calculating combined likelihood ratio (LRc), we used the product rule and multiply LR obtained with autosomal STRs with LR obtained with mtDNA. The combined likelihood ratio was 1×10^6 and the combined posterior probability of sibling (PPc) was 99.99%. The prior probability was set to 0.01 (number of victims in mass grave was 88).

The statistical analyses showed a high confidence of correct identification, since victim 1 and 2 from the Konfin I mass grave were identified with posterior probability higher than 99.9% (Biesecker et al. 2005; Brenner and Weir 2003; Prinz et al. 2007).

EXAMPLE OF GENETIC ANALYSES OF HUMAN SKELETAL REMAINS FROM ARCHAEOLOGICAL SITE

In 2009, archaeologists excavated five skeletons from the 17th century archaeological site in Ljubljana. They were found in the side chapel of the church in the Franciscan monastery that was the Auersperg tomb. Besides the skeletons, a bronze bowl with a heart was found and the name of Ferdinand II and the year of death (1655 - 1706) engraved.

In Figure 18 there are photos of skeletons excavated at the Auersperg tomb archaeological site (left) and the bronze bowl (two middle pictures) and the heart (right) found in it.

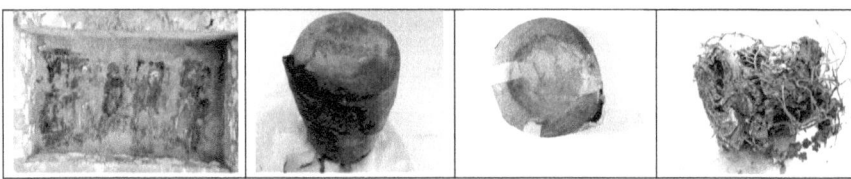

Figure 18: Photos of skeletons excavated at the Auersperg tomb archaeological site (left) (photo: Matjaž Bizjak) and the bronze bowl (two middle pictures) (photo: Matjaž Bizjak) and the heart (right) found in it (photo: Katarina Podgoršek).

The Auersperg (Turjaški) were the most influential aristocratic family on Slovenian territory and one of the richest in the Hapsburg Empire. They settled Kranjska in the 11th century and left Slovenia before the Second World War. In 2011, we were asked to identify five skeletons excavated from the Auersperg chapel. Not much of the skeletons remained and the bones were degraded to small pieces. Fragments of femurs and teeth were preserved only for two skeletons and for the rest of the three skeletons the fragments of cranium were used for molecular genetic analyses.

In Figure 19 there are photos of the skeletons excavated at the Auersperg tomb archaeological site.

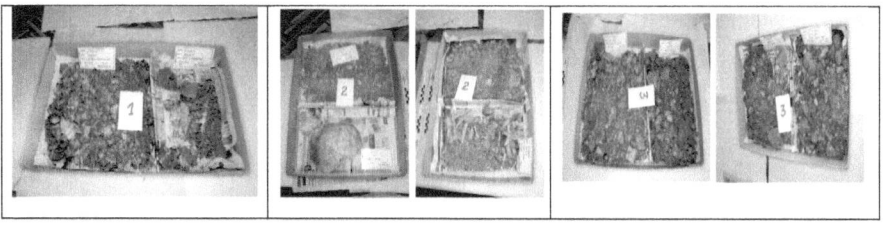

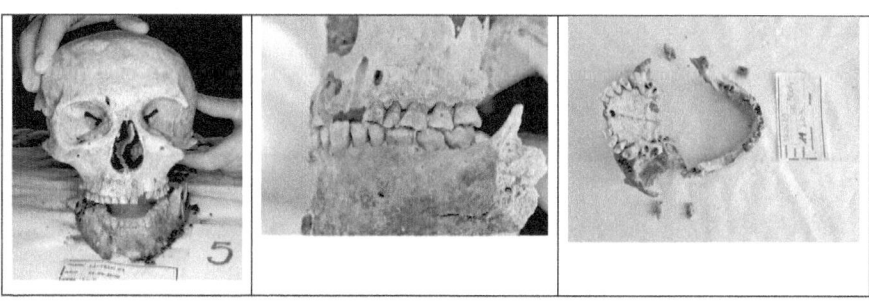

Figure 19: Photos of the skeletons excavated at the Auersperg tomb archaeological site (photo: Rudi Bevc).

Genomic DNA extraction was done at least twice from each bone and tooth. For traceability in the event of contamination, we created an elimination database including genetic profiles of the nuclear and mtDNA of all persons (archaeologists, anthropologists and geneticists) that had been in contact with the skeletal remains. We extracted up to 10.7 ng DNA/g of bone and tooth powder from the Auersperg chapel archaeological site skeletal remains. We managed to obtain nuclear DNA for successful STR typing from skeletal remains that were over 300 years old. From one skeleton, we obtained a complete male genetic profile of autosomal DNA, an almost complete Y-STR haplotype that enables us to track paternal line and an mtDNA haplotype for HVI and HVII region that enables us to track maternal line (Zupanič Pajnič et al. 2012b; Zupanič Pajnič 2013a). After comparing the profiles with the elimination database, no match was found. We are waiting for the family reference samples for comparison with genetic profiles obtained and for identification of the skeleton excavated from the Auersperg chapel archaeological site. This is the first archaeogenetic research performed in Slovenia.

In Figure 20 we present the electropherograms of complete NGM, ESSplex and MiniFiler genetic profiles obtained from the teeth of the 300 year old skeleton excavated from the Auersperg tomb archaeological site.

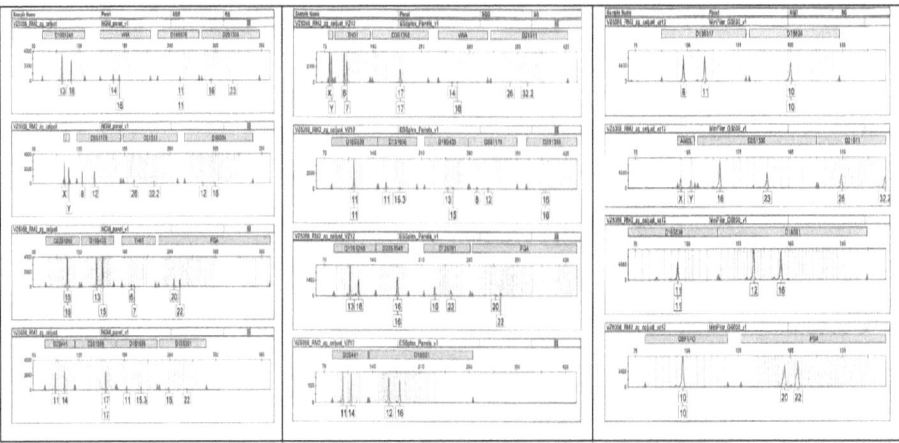

Figure 20: The electropherograms of complete NGM, ESSplex and MiniFiler genetic profiles obtained from the teeth of the 300 year old skeleton excavated from the Auersperg tomb archaeological site.

In figure 21 we present the electropherograms of the Y - chromosomal Yfiler haplotypes obtained from the teeth of the 300 year old skeleton excavated from the Auersperg tomb archaeological site.

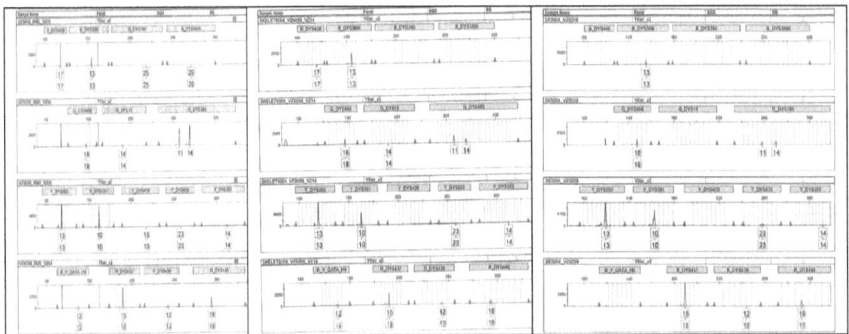

Figure 21: The electropherograms of the Y - chromosomal Yfiler haplotypes obtained from the teeth of the 300 year old skeleton excavated from the Auersperg tomb archaeological site.

CONCLUSION

We managed to obtain nuclear DNA from bones that were almost 70 years old for successful STR typing of 88 victims from the Konfin I mass grave (Zupanič Pajnič et al. 2010), three victims from the grave at Mount Storžič (Zupanič Pajnič 2008) and 25 victims from the Bodovlje Gorge (Bodoveljska grapa) mass grave (Zupanič Pajnič 2007). The methods of DNA extraction and amplification described here have proved to be highly efficient because we obtained up to 100 ng DNA/g of bones and complete genetic profiles of autosomal STRs. DNA extraction proved effective from relatively small 0.5g bone samples. The identity of the genetic profiles of bone specimens was verified by amplification of STRs with two different amplification kits. For degraded samples, we additionally used the amplification kit for mini STRs. When the genetic profiles of bones matched living relatives, the recommended PP of 99.9% was higher in all identifications, indicating that a sufficient number of genetic markers were investigated in identifying skeletal remains.

In the process of identifying victims in the Slovenian mass graves, we minimized the possibility of contamination during genetic investigations. The authenticity of genetic profiles of bones was confirmed by clean isolation and amplification-negative controls for nuclear DNA, identical genetic profiles obtained using two different autosomal STR amplification kits and mismatch of genetic profiles of bones with persons from the elimination database.

Different environmental factors, combined with underground burial for almost 70 years, detrimentally affected the ability to recover intact and uncompromised DNA from Second World War skeletal remains. Advanced extraction and purification techniques were found to be essential tools for obtaining sufficient DNA from bones and teeth uncovered from Slovenian mass graves. Extraction and purification methods using the EZ1 biorobot (Qiagen), together with more sensitive and robust new amplification kits with the ESS loci that are more tolerant of common inhibitors, allowed us to overcome the challenges associated with processing compromised skeletal remains and ultimately obtain STR DNA profiles in almost all of the bones and teeth with the commercially available autosomal STR kits.

Experiences obtained in processing and DNA typing of Second World War skeletal remains are very helpful for us in optimizing methods for molecular anthropologic analyses of much older skeletal remains from archaeological sites (Zupanič Pajnič et al. 2012b; Zupanič Pajnič 2013a).

ACKNOWLEDGEMENTS

The author gratefully acknowledges the contribution of Barbara Gornjak Pogorelc and Katja Vodopivec Mohorčič towards processing and DNA typing of bones and teeth, Barbara Gornjak Pogorelc and Petra Balažic for translation of parts of the text into English and the editorial board of the *Medicinski razgledi* medical journal for permission that parts of the monograph are based on an article originally published in the Slovenian language (Zupanič Pajnič 2013c).

LITERATURE

Alaeddini R, Walsh SJ, Abbas A. Forensic implications of genetic analysis from degraded DNA - A review. Forensic Sci Int Genet. 2010;4:149-157.

Alaeddini R. Forensic implication of PCR inhibition - A review. Forensic Sci Int Genet. 2012;6:297-305.

Albinsson L, Noren L, Hedell R. Swedish population data and concordance for the kits PowerPlex® ESX 16 System, PowerPlex® ESI 16 System, AmpFlSTR® NGMTM, AmpFlSTR® SGMTM Plus and Investigator ESSplex. Forensic Sci Int Genet. 2011;5:e89-92.

Alonso A, Andelinović Š, Martin P. DNA typing from skeletal remains: evaluation of multiplex and megaplex STR systems on DNA isolated from bone and teeth samples. Croat Med J. 2001;42:260-266.

Alonso A, Martin P, Albarran C, Garcia P, Primorac D, Garcia O, Fernandez L. Specific quantification of human genomes from low copy number DNA samples in forensic and ancient DNA studies. Croat Med J. 2003;44:273-280.

Alonso A, Martin P, Albarran C. Real-time PCR designs to estimate nuclear and mitochondrial DNA copy number in forensic and ancient DNA studies. Forensic Sci Int. 2004;139:141-149.

Alvarez Garcia A, Munoz I, Pestoni C, Lareu MV, Rodrigues-Calvo MS, Carracedo A. Effect of environmental factors on PCR-DNA analysis from dental pulp. Int J Legal Med. 1996;109:125-129.

Amory S, Huel R, Bilić A, Loreille O, Parsons TJ. Automatable full demineralization DNA extraction procedure from degraded skeletal remains. Forensic Sci Int Genet. 2012;6:398-406.

Anderson S, Bankier AT, Barrell BG. Sequence and organization of the human mitochondrial genome. Nature 1981;290:457-465.

Anderung C, Persson P, Bouwman A, Elburg R, Gotherstrom A. Fishing for ancient DNA. Forensic Sci Int Genet. 2008;2:104-7.

Andreasson H, Nilsson M, Budowle B, Lundberg H, Allen M. Nuclear and mitochondrial DNA quantification of various forensic materials. Forensic Sci Int. 2006;164:56-64.

Anslinger K, Weichhold G, Keil W, Bayer B, Eisenmenger W. Identification of the skeletal remains of Martin Bormann by mtDNA analysis. Int J Legal Med. 2001;114:194-196.

Baker LE, McCormick WF, Matteson KJ. A silica-based mitochondrial DNA extraction method applied to forensic hair shafts and teeth. J Forensic Sci. 2001;46:126-130.

Balding DJ, Nichols RA. DNA profile match probability calculation: how to allow for population stratification, relatedness, database selection and single bands. Forensic Sci Int. 1994;64:125-140.

Bandelt HJ, Parson W. Consistent treatment of length variants in the human mtDNA control region: a reappraisal. Int J Legal Med 2008;122:11-21.

Bär W, Brinkmann B, Budowle B, Carracedo A, Gill P, Holland M. DNA commission of the International Society for Forensic Genetics: guidelines for mitochondrial DNA typing. Int J Legal Med. 2000;113:193-196.

Bender K, Schneider PM, Rittner C. Application of mtDNA sequence analysis in forensic casework for the identification of human remains. Forensic Sci Int. 2000;113:103-107.

Biesecker LG, Bailey-Wilson JE, Ballantyne J, Baum H, Bieber FR, Brenner C, Budowle B, Butler JM, Carmody G, Conneally PM, Duceman B, Eisenberg A, Forman L, Kidd KK, Leclair B, Niezgoda S, Parsons TJ, Pugh E, Shaler R, Sherry ST, Sozer A, Walsh A. DNA identification after the 9/11 World Trade Center attack. Science 2005;310:1122-1123.

Bogdanowicz W, Allen M, Branicki W, Lembring M, Gajewska M, Kupiec T. Genetic identification of putative remains of the famous astronomer Nicolaus Copernicus. Proc Natl Acad Sci USA. 2009;106:12279-82.

Boles CT, Snow CC, Stover E. Forensic DNA testing on skeletal remains from mass graves: a pilot project in Guatemala. J Forensic Sci. 1995;40:349-355.

Boom R, Sol CJA, Salimans MMM, Jansen CL, Wertheim van Dillen PME, Van der Noordaa J. Rapid and simple method for purification of nucleic acids. J Clinical Microb. 1990;28:495-503.

Brenner CH, Weir BS. Issues and strategies in the DNA identification of World Trade Center victims. Theoret Popul Biol. 2003;63:173-178.

Brenner CH. DNA-VIEW 2007 user guide. 2007; Oakland

Brown TA, Brown KA. Ancient DNA and the archaeologist. Antiquity 1992;66:10-23.

Budowle B, Ge J, Chakraborty R, Eisenberg AJ, Green R, Mulero J. Population genetic analyses of the NGM STR loci. Int J Legal Med. 2010;125:101-9.

Budowle B, Smith J, Moretti T, DiZinno J. DNA typing protocols: molecular biology and forensic analysis. 2000, Washington, Eaton Publishing, p. 304.

Burger J, Hummel S, Hermann B, Henke W. DNA preservation: a microsatellite-DNA study on ancient skeletal remains. Electrophoresis 1999;20:1722-28.

Carracedo A, Bär W, Lincoln P, Mary W. DNA commission of the International Society for Forensic Genetics: guidelines for mitochondrial DNA typing. Forensic Sci Int. 2000;110:79-85.

Castella V, Dimo-Simonin N, Brandt-Casadevall C, Robinson N, Sougy M, Taroni F, Mangin P. Forensic identification of urine sample: a comparison between nuclear and mitochondrial DNA markers. Int J Legal Med. 2006;120:67-72.

Cattaneo C, Craig OE, James NT, Sokol RJ. Comparison of three DNA extraction methods on bone and blood stains up to 43 years old and amplification of three different gene sequences. J Forensic Sci. 1997;42:1126-1135.

Davoren J, Vanek D, Konjhodzić R, Crews J, Huffine E, Parsons TJ. Highly effective DNA extraction method for nuclear short tandem repeat testing of skeletal remains from mass graves. Croat Med J. 2007;48:478-485.

Edson SM, Ross JP, Coble MD, Parsons TJ, Barritt SM. Naming the dead - confronting the realities of rapid identification of degraded skeletal remains. Forensic Sci Reviews 2004;16:64-89.

Eglinton G, Logan GA. Molecular preservation. Philos Trans R Soc London Ser B 1991;333:315-27; discussion 27-28.

Ferenc M. Topografija evidentiranih grobišč (Topography of documented mass graves). In: Dežman J, ed. Poročilo Komisije vlade Republike Slovenije za reševanje vprašanj prikritih grobišč 2005-2008. Ljubljana: Družina; 2008. p. 7-27.

Fisher DL, Holland MM, Mitchell L, Sledzik PS, Wilcox AW, Wadhams M, Weedn VW. Extraction evaluation and amplification of DNA from decalcified and undecalcified United States civil war bone. J Forensic Sci. 1993;38:60-68.

Garcia O, Alonso J, Cano JA, Garcia R, Luque GM, de Yuso M, Maulini S, Parra D, Yurrebaso I. Population genetic data and concordance study for the kits Identifiler, NGM, powerPlex ESX 17 System and Investigator Essplex in Spain. Forensic Sci Int Genet. 2012;6:e78-e79.

Gaytmenn R, Sweet D. Quantification of forensic DNA from various regions of human teeth. J Forensic Sci. 2003;48:622-625.

Gilbert MTP, Rudbeck L, Willerslev E. Biochemical and physical correlates of DNA contamination in archaeological bone and teeth excavated at Matera, Italy. J Archaeol Sci. 2005;32:785-93.

Gill CR, Duewer DL, Kline MC, Sprecher CJ, McLaren RS, Rabbach DR. Concordance and population studies along with stutter and peak height ratio analysis for the PowerPlex® ESX 17 and ESI 17 System. Forensic Sci Int Genet. 2011;5:269-75.

Gill P, Fereday L, Morling N, Schneider PM. New multiplexes for Europe Amendments and clarification of strategic development. Forensic Sci Int. 2006b;163:155-7.

Gill P, Fereday L, Morling N, Schneider PM. The evolution of DNA databases-recommendations for new European loci. Forensic Sci Int. 2006a;156:242-4.

Graham EAM. DNA reviews: Ancient DNA. Forensic Sci Med Patholol. 2007;3:221-225.

Gusmao L, Butler JM, Carracedo A, Gill P, Kayser M, Mayr WR, Morling N, Prinz M, Roewer L, Tyler-Smith C, Schneider PM. DNA commission of the international society of forensic genetics (ISFG): an update of the recommendations on the use of Y STRs in the forensic analysis. Int J Legal Med. 2006;120:191-200.

Hagelberg E, Bell LS, Allen T, Boyde A, Jones SJ, Cledd JB. Analyses of ancient bone DNA: techniques and applications. Philos Trans R Soc Lond B Biol Sci. 1991;333:399-407.

Hagelberg E, Gray IC, Jeffreys AJ. Identification of the skeletal remains of a murder victim by DNA analysis. Nature 1991;352:427-429.

Hagelberg E, Sykes B, Hedges R. Ancient bone DNA amplified. Nature 1989;342:485.

Handt O, Krings M, Ward RH, Pääbo S. The retrieval of ancient human DNA sequences. Am J Hum Genet. 1991;59:368-76.

Handt O, Richards M, Trommsdorf M, Kilger C, Simanainen J. Molecular genetic analyses of the Tyrolean Ice Man. Science 1994;264:1775-78.

Hill CR, Kline MC, Duewer DL, Butler JM. Concordance testing comparing STR multiplex kits with a standard data set. Forensic Sci Int: Genetic Supplement Series. 2011;3:e188-e189.

Hochmeister MN, Budowle B, Borer V, Rudin O, Bohnert M, Dirnhofer R. Confirmation of the identity of human skeletal remains using multiplex PCR amplification and typing kitas. J Forensic Sci. 1995;40:701-705.

Hochmeister MN, Budowle MD, Borer UV, Eggmann U, Comey CT, Dirnhofer R. Typing of deoxyribonucleic acid (DNA) extracted from compact bone from human remains. J Forensic Sci. 1991;36:1649-1661.

Hofreiter M, Serre D, Poinar HN, Kuch M, Pääbo S. Ancient DNA. Nat Rev Genet. 2001;2:353-59.

Hopwood AJ, Mannucci A, Sullivan KM. DNA typing from human faeces. Int J Legal Med. 1996;108:237-243.

Höss M, Jaruga P, Zastawny TH, Dizdaroglu M, Pääbo S. DNA damage and DNA sequence retrieval from ancient tissues. Nucleic Acids Res. 1996;24:1304-1307.

Irwin JA, Edson SM, Loreille O, Just RS, Barritt SM, Lee DA, Holland TD, Parsons TJ, Leney MD. DNA identification of "Earthquake McGoon" 50 years postmortem. J Forensic Sci. 2007a;52:1115-1118.

Irwin JA, Just RS, Loreille OM, Parsons TJ. Characterization of a modified amplification approach for improved STR recovery from severely degraded skeletal elements. Forensic Sci Int Genet. 2012;6:578-587.

Irwin JA, Leney MD, Loreille O, Barritt SM, Christensen AF, Holland TD, Smith BC, Parsons TJ. Application of low copy number STR typing to the identification of aged, degraded skeletal remains. J Forensic Sci. 2007b;52:1322-1327.

Jakubowska J, Maciejewska A, Pawlowski R. Comparison of three methods of DNA extraction from human bones with different degrees of degradation. Int J Legal Med. 2012;126:173-178.

Jamnik P. Ugotavljanje identitete žrtev iz brezna pri Konfinu I v arhivskih virih. (Identifying victims in the Konfin I mass grave using the documents from the archives). In: Dežman J, ed. Poročilo Komisije vlade Republike Slovenije za reševanje vprašanj prikritih grobišč 2005-2008. Ljubljana: Družina; 2008. p. 83-93.

Jehaes E. Mitochondrial DNA analysis on remains of a putative son of Louis XVI, King of France and Marie-Antoinette. Europ J Hum Gen. 1998;6:383-395.

Kalmar T, Bachrati CZ, Marcsik A, Rasko I. A simple and efficient method for PCR amplifiable DNA extraction from ancient bones. Nucleic Acids Res. 2000;28:e67.

Kemp BM, Smith DG. Use of bleach to eliminate contaminating DNA from the surface of bones and teeth. Forensic Sci Int. 2005;154:53-61.

Kishore R, Hardy WR, Anderson VJ, Sanchez NA, Buoncristiani MR. Optimization of DNA extraction from low-yield and degraded samples using the biorobot EZ1 and biorobot M48. J Forensic Sci. 2006;51:1055-1061.

Kolmann CJ, Tuross N. Ancient DNA analysis of human populations. Am J Phys Anthropol. 2000;111:5-23.

Koyama H, Iwasa M, Ohtani S, Ohira H, Tsuchimochi T. Personal identification from human remains by mitochondrial DNA sequencing. Am J Forensic Med Pathol. 2002;23:272-276.

Kutranov S. Comparison study of four different 16-locus "expanded ESS" STR kits. Forensic Sci Int: Genetic Supplement Series. 2011;3:e196-e197.

Lee EJ, Luedtke JG, Allison JL, Arber CE, Merriwether DA, Steadman DW. The effects of different maceration techniques on nuclear DNA amplification using human bone. J Forensic Sci. 2010b;55:1032-38.

Lee H, Pagliaro E, Berka K, Folk N, Anderson D, Ruano G, Keith T, Phipps P, Herrin G, Garner D, Gaensslen R. Genetic markers in human bone: I. deoxyribonucleic acid (DNA) analysis. J Forensic Sci. 1991;36:320-330.

Lee HY, Kim NY, Park MJ, Sim JE, Yang Wi, Shin KJ. DNA typing for the identification of old skeletal remains from Korean war victims. J Forensic Sci. 2010c;55:1422-1429.

Lee HY, Park MJ, Kim NY, Sim JE, Yang WI, Shin KJ. Simple and highly effective DNA extraction method from old skeletal remains using silica columns. Forensic Sci Int Genet. 2010a;4:275-80.

Lindahl T. Instability and decay of the primary structure of DNA. Nature 1993;362:709-15.

Lleonart R, Rirgo E, Sainz de la Pena MV, Bacallao K, Amaro F. Forensic identification of skeletal remains from members of Ernesto Che Guevara's guerrillas in Bolivia based on DNA typing. Int J Legal Med 2000;113:98-101.
Loreille OM, Diegoli TM, Irwin JA, Coble MD, Parsons TJ. High efficiency DNA extraction from bone by total demineralization. Forensic Sci Int Genet. 2007;1(2):91-195.

Lutz S, Weisser HJ, Heizmann J, Pollak S. MTDNA as a tool for identification of human remains - Identification using mtDNA. Int J Legal Med. 1996;109:205-209.

Malmstrom H. Ancient DNA as a means to investigate the European Neolithic. Doctoral dissertation, 2007, Uppsala University, Uppsala, Sweden

Miloš A, Selmanović A, Smajlović L, Huel RLM, Katzmarzyk C, Rizvić A, Parsons JP. Success rates of nuclear short tandem repeat typing from different skeletal elements. Croat Med J. 2007;48:486-93.

Misner LM, Halvorson AC, Dreier JL, Ubelaker DH, Foran DR. The correlation between skeletal weathering and DNA quality and quantity. J Forensic Sci. 2009;54:822-828.

Molnar A, Zalan A, Horvath G, Pamjav H. Allele distribution of the new European Standard Set (ESS) loci in the Hungarian population. Forensic Sci Int Genet. 2011;5:555-6.

Montpetit SA, Fitch IT, O'Donnell PT. A simple automated instrument for DNA extraction in forensic casework. J Forensic Sci. 2005;50:555-563.

Mörnstad H, Pfeiffer H, Yoon C, Teivens A. Demonstration and semi-quantification of mtDNA from human dentine and its relation to age. Int. J Legal Med. 1999;112:98-100.

Nagy M, Otremba P, Krüger C, Bergner-Greiner S, Anders P, Henske B, Prinz M, Roewer L. Optimization and validation of a fully automated silica-coated magnetic beads technology in forensics. Forensic Sci Int. 2005;152:13-22.

Noonan JP, Coop G, Kudaravalli S, Smith D, Krause J. Sequencing and analysis of Neanderthal genomic DNA. Science 2006;314:1113-1118.

Ohira H, Yamada Y. Advantages of dental mitochondrial DNA from detection and classification of the sequence variation using restriction fragment length polymorphisms. Am J Forensic Med Pathol. 1999;20:261-268.

Pääbo S, Poinar H, Serre D. Genetic analyses from ancient DNA. Annu Rev Genet. 2004;38:645-79.

Pääbo S. Amplifying ancient DNA. In PCR-Protocols and Amplifications-A Laboratory Manual, ed. MA Innis, DH Gelfand, JJ Sninsky, TJ White, 1990, pp 159-66. San Diego: Academic

Pääbo S. Ancient DNA: extraction, characterization, molecular cloning, and enzymatic amplification. Proc Natl Acad Sci USA 1989;86:1939-43.

Palo JU, Hedman M, Soderholm N, Sajantila A. Repatriation and identification of Finnish World War II soldiers. Croat Med J. 2007;48:528-535.

Parson W, Dür A. EMPOP - A forensic mtDNA database. Forensic Sci Int: Genetics 2007;1:88-92.

Parson W, Parsons TJ, Scheithauer R, Holland MM. Population data for 101 Austrian Caucasian mitochondrial DNA d-loop sequences: Application of mDNA sequence analysis to a forensic case. Int J Legal Med. 1998;111:124-132.

Parys-Proszek A, Kupiec T, Wolanska-Nowak P, Branicki W. Genetic variation of 15 autosomal STR loci in a population sample from Poland. Leg Med. 2010;12:246-8.

Petrovič D, Zorc M. Histologija. Ljubljana, Univerza v Ljubljani Medicinska fakulteta Inštitut za histologijo in embriologijo 2005, p. 35-42, p.115-123, p.166-168.

Pfeiffer H, Steighner R, Fisher R, Mörnstad H, Yoon CL, Holland MM. Mitochondrial DNA extraction and typing from isolated dentin-experimental evaluation in Korean population. Int J Legal Med. 1998;111:309-313.

Pfeiffer H. Hühne J, Seitz B, Brinkmann B. Influence of soil storage and exposure period on DNA recovery from teeth. Int J Legal Med. 1999;112:142-144.

Poetsch M, Bayer K, Ergin Z, Milbrath M, Schwark T, von Wurmb-Schwark N. First experiences using the new PowerPlex® ESX 17 and ESI17 kits in casework analysis and allele frequencies for two different regions in Germany. Int J Legal Med. 2011;125:733-9.

Poetsch M, Ergin Z, Bayer K, El-Mostaqim D, Rokotomavo N, Browne ENL. The new PowerPlex® ESX 17 and ESI17 kits in paternity and maternity analyses involving people from Africaincluding allele frequencies for three African populations. Int J Legal Med. 2010;125:149-54.

Poinar HN, Hoss M, Bada JL, Pääbo S. Amino acid racemisation and the preservation of ancient DNA. Science 1996;272:864-866.

Previdere C, Grignani P, Presciuttini S. Italian population data for the new ENFSI/EDNAP loci D1S1656, D2S441, D10S1248, D12S391, D22S1045. The GeFI collaborative exercise and concordance study. Forensic Sci Int: Genetic Supplement Series. 2011;3:e238-e239.

Prince AM, Andrus L. PCR how to kill unwanted DNA. Biotechniques 1992;12:358-360.

Prinz M, Carracedo A, Mayr WR, Morling N, Parsons TJ, Sajantila A, Scheithauer R, Schmitter H, Schneider PM. DNA Commission of the International Society for Forensic Genetics (ISFG): Recommendations regarding the role of forensic genetics for disaster victims identification (DVI). Forensic Sci Int Genet. 2007;1:3-12.

Pruvost M, Schwarz R, Correia VB. Freshly excavated fossil bones are best for amplification of ancient DNA. Proc Natl Acad Sci USA 2007;104:739-744.
Putkonen MT, Palo JU, Cano JM, Hedman M, Sajantila A. Factors affecting the STR amplification sucess in poorly preserved bone samples. Investigative genetics 2010;1:9.

Rennick SL, Fenton TW, Foran DR. The effects of skeletal preparation techniques on DNA from human and non-human bone. J Forensic Sci. 2005;50:1016-1019.

Salamon M, Tuross N, Arensburg B, Weiner S. Relatively well preserved DNA is present in the crystal aggregates of fossil bones. Proc Natl Acad Sci USA 2005;102:13783-13788.

Sampierto ML, Gilbert MT, Lao O, Caramelli D, Lari M, Berranpetit J, Lalueza-Fox C. Tracking down human contamination in ancient human teeth. Mol Biol Evol. 2006;23:1801-1807.

Scherer M, Muller D, Begemann S, Steeger B, Pakulla S, Breitbach M, Cornelius S, Bochmann L, Prochnow A, Schnibbe T, Engel H. The Investigator Essplex Plus Kit: Fast, sensitive, and robust amplification of the European Standard Set of loci. Forensic Sci Int: Genetic Supplement Series. 2011;3:e407-e408.

Schwartz TR, Schwartz EA, Mieszerski L, McNally L, Kobilinsky L. Characterization of deoxyribonucleic acid (DNA) obtained from teeth subjected to various environmental conditions. J Forensic Sci. 1991;36:979-990.

Seo SB, Lee HY, Zhang AH, Kim HY, Shin DH, Lee SD. Effects of humic acid on DNA quantification with Quantifiler Human DNA Quantification kit and short tandem repeat amplification efficiency. Int J Legal Med. 2012;126:961-968.

Seo Y, Uchiyama T, Shimizu K, Takahama K. Identification of remains by sequencing of mitochondrial DNA control region. Am J Forensic Med Pathol. 2000;20:138-143.

Shaw K, Sesardić I, Bristol N, Ames C, Dagnall K, Ellis C, Whittaker F, Daniel B. Comparison of the effects of sterilisation techniques on subsequent DNA profiling. Int J Legal Med. 2008;122:29-33.

Smith BC, Fisher DL, Weedn VW, Warnock GR, Holland MM. A systematic approach to the sampling of dental DNA. J Forensic Sci. 1993;38:1194-1209.

Smith CI, Chamberlain AT, Riley MS, Cooper A, Stringer CB, Collins MJ. Neanderthal DNA. Not just old but old and cold? Nature 2001;410:771-772.

Smith CI, Chamberlain AT, Riley MS, Stringer C, Collins MJ. The thermal history of human fossils and the likelihood of successful DNA amplification. J Hum Evol. 2003;45:203-217.

Sprecher CJ, McLaren RS, Rabbach D, Krenke B, Ensenberger MG, Fulmer PM. PowerPlex® ESX and ESI Systems: A suite of new STR systems designed to meet the changing needs of the DNAtyping community. Forensic Sci International Genet Suppl Ser. 2009;2:2-4.

Stone AC, Starrs JE, Stoneking M. Mitochondrial DNA analysis of the presumptive remains of Jesse James. J Forensic Sci. 2001;46:173-176.

Sullivan KM, Hopgood R, Gill P. Identification of human remains by amplification and automated sequencing of mitochondrial DNA. Int J Legal Med. 1992;105:83-86.

Sweet D, Hildebrand D, Phillips D. Identification of skeleton using DNA from teeth and a PAP smear. J Forensic Sci. 1999;44:630-633.

Sweet D, Hildebrand D. Recovery of DNA from human teeth by cryogenic grinding. J Forensic Sci. 1998;43:1199-1202.

Sweet DJ, Sweet CHW. DNA analysis of dental pulp to link incinerated remains of homicide victim to crime scene. J Forensic Sci. 1995;40:310-314.

Tahir MA, Balraj E, Luke L, Gilbert T, Hamby JE, Amjad M. DNA typing of samples for polymarker DQA1 and nine STR loci from human body exhumed after 27 years. J Forensic Sci. 2000;45:902-907.

Tamariz J, Voynarovska K, Prinz M, Caragine T. The application of ultraviolet irradiation to exogenous sources of DNA in plasticware and water for the amplification of low copy number DNA. J Forensic Sci. 2006;51:790-794.

Tsuchimochi T, Iwasa M, Maeno Y, Koyama H, Inoue H, Isobe I, Matoba R, Yokoi M, Nagao M. Chelating resin-based extraction of DNA from dental pulp and sex determination from incinerated teeth with Y-chromosomal alphoid repeat and short tandem repeats. Am J Forensic Med Pathol. 2002;23:268-271.

Tucker VC, Hopwood AJ, Sprecher CJ, McLaren RS, Rabbach DR, Ensenberger MG. Developmental validation of the PowerPlex® ESI 16 and PowerPlex® ESI 17 Systems: STR multiplex for the new European standard. Forensic Sci Int Genet. 2011;5:436-48.

Tully G, Bär W, Brinkmann B, Carracedo A, Gill P, Morling N. Considerations by the European DNA profiling (EDNAP) group on the working practices nomenclature and interpretation of mitochondrial DNA profiles. Forensic Sci Int 2001;124:83-91.

Tuross N. The biochemistry of ancient DNA in bone. Experientia 1994;50:530-535.

Valdiosera C, Garcia N, Dalen L, Smith C. Typing single polymorphic nucleotides in mitochondrial DNA as a way to access Middle Pleistocene DNA. Biol Lett. 2006;2:601-603.

Valgren C, Wester S, Hansson O. A comparison of three automated DNA purification methods in forensic casework. Forensic Sci Int: Genetics Supplement Series 2008;1:76-77.

Vanek D, Saskova L, Koch H. Kinship and Y-chromosome analysis of 7th century human remains: Novel DNA extraction and typing procedure for ancient material. Croat Med J. 2009;50:286-295.

Walsh B, Redd AJ, Hammer MF. Joint match probabilities for Y chromosomal and autosomal markers. Forensic Sci Int. 2008;174:234-238.

Wandeler P, Smith S, Morin PA, Pettifor RA, Funk SM. Paterns of nuclear DNA degradation over time-a case study in historic teeth samples. Mol Ecol. 2003;12:1087-93.

Welch LA, Gill P, Phillips C, Ansell R, Morling N, Parson W, Palo JU, Bastisch I. European Network of forensic Science Institutes (ENFSI): Evaluation of new commercial STR multiplexes that include the European Standard Set (ESS) of markers. Forensic Sci Int Genet. 2012;6:819-826.

Willuweit S, Roewer L. Y chromosome haplotype reference database (YHRD): Update. Forensic Sci Int: Genetics 2007;1:83-87.

Wilson MR, DiZinno JA, Polanskey D, Replogle J, Budowle B. Validation of mitochondrial DNA sequencing for forensic casework analysis. Int J Legal Med. 1995;108:68-74.

Yamomoto T, Uchihi R, Kojima T, Nozawa H, Huang XL, Tamaki K, Katsumata Y. Maternal identification from skeletal remains of an infant kept by the alleged mother for 16 years with DNA typing. J Forensic Sci. 1998;43:701-705.

Yurrebaso I, Ajuriagerra JA, Alday A, Lezama I, Perez JA, Romon E. Allele frequencies and concordance study between the Identifiler and the ESX 17 System in the Basque Country population. Forensic Sci Int Genet. 2011;5:79-80.

Zupanc T, Balažic J, Štefanič B, Zupanič Pajnič I. Performance of the Human Quantifiler, the Investigator Quantiplex, and the Investigator ESSplex Plus kit for quantification and nuclear DNA typing of old skeletal remains. Rom J Legal Med 2013;21(2):119-124.

Zupanič I, Balažic J, Komel R. Analysis of nine short tandem repeat (STR) loci in the Slovenian population. Int J Legal Med 1998;111:248-250.

Zupanič Pajnič I, Balažic J, Komel R. Sequence polymorphism of the mitochondrial DNA control region in the Slovenian population. Int J Legal Med. 2004;118:1-4.

Zupanič Pajnič I, Gornjak Pogorelc B, Balažic J, Zupanc T, Štefanič B. Highly efficient nuclear DNA typing of the World War II skeletal remains using three new autosomal short tandem repeat amplification kits with the extended European Standard Set of loci. Croat Med J. 2012a; 53(1):17-23.

Zupanič Pajnič I, Pogorelc BG, Balažic J, Horvat M. Molecular genetic analyses of skeleton excavated from Auersperg Chapel archaeological site in Slovenia.

In: DNA in forensics 2012: "exploring the phylogenies". Abstract book and conference program of the 5th EMPOP meeting, 8th Y-chromosomal user workshop; 2012 Sep 6-8; Innsbruck. Innsbruck: 2012b; 108.

Zupanič Pajnič I, Šterlinko H, Balažic J, Komel R. Parentage testing with 14 STR loci and population data for 5 STRs in the Slovenian population. Int J Legal Med 2001;114:178-180.

Zupanič Pajnič I. A comparative analysis of the AmpF/STR Identifiler and PowerPlex 16 autosomal Short Tandem Repeat (STR) amplification kits on the skeletal remains excavated from Second World War mass graves in Slovenia. Rom J Legal Med 2013b;21(1):73-78.

Zupanič Pajnič I. Forenzična genetika (Forensic Genetics). Med Razgl. 2011b;50(3):325-340.

Zupanič Pajnič I. Identifikacija oseb iz starih in slabo ohranjenih bioloških materialov s polimorfizmi mitohondrijske DNA (Human identification from old and badly preserved biological material with polymorphisms of mitochondrial DNA (doctoral dissertation). Ljubljana: Univerza v Ljubljani; 2007.

Zupanič Pajnič I. Molekularno genetska identifikacija domobranskih žrtev (Molecular genetic identification of Slovenian home guard victims). Zdrav Vestn. 2008;77(11):745-50.

Zupanič Pajnič I. Molekularnogenetska identifikacija skeletnih ostankov (Molecular genetic identification of skeletal remains). Med Razgl. 2013c;52:213-234.

Zupanič Pajnič I. Molekularnogenetska preiskava 300 let starih skeletov iz Auerspergove grobnice (Molecular genetic analyses of 300 years old skeletons from Auersperg tomb). Zdrav Vestn. 2013a; in press.

Zupanič Pajnič I. Visoko učinkovita metoda ekstrakcije DNA iz skeletnih ostankov (Highly efficient DNA extraction method from skeletal remains). Zdrav Vestn. 2011a;80:171-181.

Zupanič-Pajnič I, Gornjak-Pogorelc B, Balažic J. Molecular genetic identification of skeletal remains from the Second world war Konfin I mass grave in Slovenia. Int J Legal Med. 2010;124(4):307-17.

i want morebooks!

Buy your books fast and straightforward online - at one of world's fastest growing online book stores! Environmentally sound due to Print-on-Demand technologies.

Buy your books online at
www.get-morebooks.com

Kaufen Sie Ihre Bücher schnell und unkompliziert online – auf einer der am schnellsten wachsenden Buchhandelsplattformen weltweit! Dank Print-On-Demand umwelt- und ressourcenschonend produziert.

Bücher schneller online kaufen
www.morebooks.de

VDM Verlagsservicegesellschaft mbH
Heinrich-Böcking-Str. 6-8   Telefon: +49 681 3720 174   info@vdm-vsg.de
D - 66121 Saarbrücken   Telefax: +49 681 3720 1749   www.vdm-vsg.de

www.ingramcontent.com/pod-product-compliance
Lightning Source LLC
Chambersburg PA
CBHW020455220526
45464CB00002B/997